中国竹工艺

第 2 版

CHINESE BAMBOO HANDICRAFTS

张齐生　程渭山　主编

中国林业出版社

图书在版编目(CIP)数据

中国竹工艺/张齐生,程渭山主编.-2版.-北京:中国林业出版社,2003.5
ISBN 7-5038-3333-5

Ⅰ.中... Ⅱ.①张... ②程... Ⅲ.竹制品-中国-摄影集 Ⅳ.TS959.2

中国版本图书馆 CIP 数据核字（2002）第 106193 号

责任编辑：杨长峰　徐小英
装帧设计：聂崇文
出版：中国林业出版社（100009 北京西城区刘海胡同 7 号）
E-mail:cfphz@public.bta.net.cn　电话：66184477
发行：新华书店北京发行所
印刷：东莞金杯印刷有限公司
版次：1997 年 6 月第 1 版（共印 1 次）
　　　2003 年 5 月第 2 版
印次：2003 年 5 月第 1 次
开本：889mm × 1194mm　1/16
印张：11
彩色照片：约 700 幅
印数：1～3000 册
定价：220.00 元

《中国竹工艺(第2版)》编委会

主　　编　　张齐生　程渭山

副 主 编　　徐华铛　吴　博　朱石麟　沈　璇　朱惠明　马广仁　王树东
　　　　　　肖江华　胡正坚　桂田　茂(日)　　辉朝茂　胡计华

编　　委　　(按姓氏笔画为序)
　　　　　　马广仁　马有基　王　正　王树东　朱石麟　朱惠明　乔海清
　　　　　　吴　博　张齐生　张祖湘　张新萍　肖江华　沈　璇　周仁龙
　　　　　　周吉仲　胡正坚　胡计华　桂田　茂(日)　　徐华铛　程渭山
　　　　　　辉朝茂

主要摄影　　马有基　余　磊
其他摄影及照片提供者
　　　　　　徐华铛　辉朝茂　陈　林　任绳武　李　华　韩　祥　张建国
　　　　　　孙万愚　郭伯南　徐　燕　徐积锋　刘慎辉　王　健　刘嘉峰
　　　　　　史舟棠　余文忠　林　欣　张培新　邹大勇　袁志稳　丁令嗣
　　　　　　何福礼　曾伟人　杨乃燕　钱忠苗　张德和　谭南森　刘徐来
　　　　　　杨宇明　刘家柱　薛嘉榕

CHIEF EDITORS: Zhang Qisheng, Cheng Weishan
DEPUTY CHIEF EDITORS:
　　Xu Huadang, Wu Bo, Zhu Shilin, Shen Xuan, Zhu Huiming, Ma Guangren,
　　Wang Shudong, Xiao Jianghua, Hu Zhengjian, Shijiru Katsurada(Japan),
　　Hui Chaomao, Hu Jihua
EDITORIAL BOARD(in alphabetical sequence):
　　Cheng Weishan, Hu Jihua, Hu Zhengjian, Hui Chaomao, Ma Guangren,
　　Ma Youji, Qiao Haiqing, Shen Xuan, Shijiru Katsurada(Japan), Wang
　　Shudong, Wang Zheng, Wu Bo, Xiao Jianghua, Xu Huadang, Zhang
　　Qisheng, Zhang Xinping, Zhang Zuxiang, Zhou Jizhong,Zhou Renlong,
　　Zhu Huiming, Zhu Shilin
PHOTOGRAPHER:Ma Youji, Yu Lei et al.

工藝

邵華澤題

序

在《中国竹工艺》即将完稿，准备付梓之际，主编要我为该画册作序，我欣然应允。当我翻阅了画册的样稿，深深为一幅幅精彩的画面所吸引，那高超的技艺和完美的艺术造型，令人惊奇、赞叹。

中国竹类资源极为丰富，又是世界上最早利用竹子的国家。几千年来，中华民族对竹子一直怀有特殊的感情，人们爱竹、咏竹、画竹、用竹，世代相传，日臻完美，形成了中国特有的竹文化传统。中国竹工艺艺术在漫长的历史发展过程中，经过历代艺人工匠创造性劳动，形成了一种独特的专门艺术。一根竹篾经过能工巧匠的手，能编织出形态各异、出神入化的各种工艺品和艺术品，一把刻刀，游刃于竹筒、竹根、竹片之上，能雕刻成栩栩如生、呼之欲出的人物形象和自然景观。人民创造了艺术，艺术陶冶了人们的情操。中国竹工艺不仅融汇了中华民族特有的传统哲理和民族风格，而且源于生活，高于生活，给人以美的想象和启迪。其绝妙之处在于"似与非似之间"；其艺术精华体现于真、善、美的高度统一。

一件好的艺术作品往往是现实主义与浪漫主义完善结合的产物，同时，又是一种生活的教科书。中国竹工艺正是以一种独特的艺术表现形式，追求这种高尚的艺术境界。它无愧于艺术百花园中一朵绚丽的花朵，也是对世界艺术殿堂的贡献。

值得一提的是，这本画册既展示了中国的竹制工艺品和艺术品的精华，又介绍了中国竹材工业化利用的最新研究成果，图文并茂、雅俗共赏，具有较强的知识性、趣味性和可读性。对读者加深中国竹工艺艺术的了解和提高欣赏能力很有帮助。

《中国竹工艺》主编和他的同事们，怀着对竹子的无限深情和对中国传统文化的热爱，为弘扬中国竹工艺艺术，振奋民族精神，促进社会主义物质文明和精神文明建设，做了一件很有意义的事情。他们深入南方林区竹乡的生产单位，广泛调查，搜集资料，现场拍摄，精心编辑，为之付出了艰辛的劳动，使画册得以与广大读者见面。我相信，在改革开放、科教兴国的伟大实践中，中国竹工艺——这颗东方艺术之明珠，必将闪烁出更加璀灿夺目的光辉。

江泽慧

1996年9月

FOREWORD

The album CHINESE BAMBOO HANDICRAFTS is being finalized and sent to the press, upon the request of the editor-in-chief I take pleasure in contributing this foreword.Having reviewed the draft of this book,I am deeply attracted by the beautiful pictures.The excellent artistic expression and perfect modelling are worthy of praise and admiration.

The bamboo resources in China are extremely abundant,the utilization of bamboo can be traced back to ancient times.Chinese people have deep feelings for bamboo. For thousands of years people love bamboo,praise it,paint it and use it, such a tradition descends generation from generation,creating a specific Chinese bamboo culture.Generations of Chinese bamboo processing masters,through persistent inventive exertion for numerous years,have developed a unique bamboo processing art.Simple bamboo stripes and threads,through the magic hands of these masters, can be converted into various mystic handicrafts;a plain carving knife,through the magic hands of these masters,can create animated images of human figures and natural landscapes on bamboo culms,roots and pieces.People develop art,while the art purifies the mind of people.Chinese bamboo processing art combines the traditional philosophical essence with national life style,it takes root from the daily life,and expresses more distinctive existence than that in daily life,offering aesthetic enlightenment and imagination.The originality of bamboo processing art lies between verisimilitude and unlikeness,in the complete harmony of truth,goodness and beauty.

Fine pieces of artistic work are always a successful combination of realism withromanticism,meanwhile,they are also an encyclopedia of life.Bamboo processing art in China is such a combination,pursuing the noble artistic realm.It fully deserves the title of the exotic flower in the garden of arts,it is also a contribution to world temple of curiosities.

It should be mentioned that this album displays the cream of Chinese bamboo handicrafts on one hand,and demonstrates the latest research results of Chinese bamboo experts on the other.The pictures and texts are excellent, they suit both fine and popular tastes.Furthermore,this book is quite informative,fascinating and readable.It is of much help to readers in understanding and appreciating bamboo art.

The editor-in-chief and his colleagues,filled with boundless enthusiasm and great devotion to our traditional culture,have done a significant job,they made studious efforts to enhance bamboo processing art,inspire national spirit,promote the construction of socialist material and spiritual civilization.They went right to productive units of bamboo growing areas in southern provinces for detailed investigation,data collection and photo-taking.Carefully edited this book is to be released as a result of continuous labor.I am convinced that the bamboo processing art— a pearl of oriental art will be more flourishing in the great undertaking of Chinese people to carry out the reform and open policy,to make our country prosperous by means of science and education.

Jiang Zehui
President Chinese Academy of Forestry
September,1996

再版前言

《中国竹工艺》画册于1997年6月出版后,得到海内外读者的热烈欢迎,第一版画册现已告罄。许多读者来信来电,希望能得到再版的画册。一些致力于竹工艺开发的专业人员在本书出版后又创作了不少高品位的竹制艺术品,有些还在海内外的各种展览会和评比会上获奖。他们不但把作品的照片寄给我们,有的还请我们去实地鉴定,希望能把这些作品通过画册介绍给广大读者。

有鉴于此,主编张齐生先生和程渭山先生在杭州召集部分编委,与中国林业出版社的徐小英先生和杨长峰先生开会共同商讨相关事宜,决定再版《中国竹工艺》;决定修改部分内容;决定增加关于丛生竹的内容,并邀请西南林学院辉朝茂先生等担任副主编。

根据会议的决定,对画册进行了全面的调整,删去了一些与竹工艺联系不紧密的照片和文字,增加了竹编织图案、竹林风光内容和一些有代表性的竹工艺品,封面也作了调整。由徐华铠副主编充实并校正了全书的照片与文字,朱石麟副主编负责全书的翻译工作。

尽管我们做了巨大努力,但由于水平有限,时间不足,这本再版的画册必然会有种种缺陷,特请读者不吝赐教。

《中国竹工艺》编委会
2002年12月

PREFACE TO SECOND EDITION

    The album CHINESE BAMBOO HANDICRAFTS, published in June 1997, was met warmly by readers home and abroad. All copies of this album were exhausted rapidly. Many readers wrote and phoned us, wishing to have the second edition. Certain professionals created new models of bamboo handicrafts of high quality after the publishing of our album, some of them were displayed and awarded prizes on various exhibitions and appraisal meetings. They sent us the photos of their creations and invited us to evaluate the new crafts, desiring to include these articles into our album.

    Encouraged by the support of readers and professionals, the editors-in-chief, Mr. Zhang Qisheng and Mr. Cheng Weishan called part of the editorial staff and hold a meeting together with Mr. Xu Xiaoying and Mr. Yang Changfeng from Chinese Forestry Publishing House. We discussed all the relevant problems and decided to prepare and release a second edition of the album; to revise part of the content of the album; to add the information concerning sympodial bamboo species and to appoint Mr. Hui Chaomao from Southwest Forestry Institute to be a vice editor-in-chief.

    In accordance with the above-mentioned-decision, the content of the album was revised; some of the photos and explanations not closely related to bamboo handicraft technology were removed; some bamboo weaving patterns, bamboo scenery pictures and photos of excellent bamboo handicrafts were added; all the photos and writings were revised carefully. The picture on first cover was also changed. The concrete activities were carried out by vice editor-in-chief Mr. Xu Huadang, and the whole text was translated into English by vice editor-in-chief Mr. Zhu Shilin.

    Though we tried our best, this album may have many shortcomings due to our insufficient capability, we whole heartedly hope readers give us comments or advice.

Editorial Committee
CHINESE BAMBOO HANDICRAFTS
December, 2002

前　言

　　中国的竹类资源十分丰富，约有30属400余种，可以说，中国是世界上竹类品种最多，产量最大的国家。这些秀丽、苍翠的竹子，丛丛相连地挺拔在群山众谷，湖畔江岸，把中国的山河装扮得多姿多采。

　　千百年来，竹子在中国文化、艺术和人们日常生活中一直闪耀着奇异的光彩，成为中国传统文化的组成部分。而最能体现竹子价值的当数竹的工艺品。"此艺与竹化，无穷出清新"。中国的竹子艺术家们，正沿着竹子本身的气质和特性，通过自己的慧眼和巧手，创造了一系列的竹子工艺品。从精雅细巧的竹子编织到天然质朴的竹子装饰；从巧夺天工的竹筒、竹节造型到鬼斧神工的竹根雕刻，无不体现竹子清雅朴实的材质美，给人以形象的启迪和美的享受。

　　"未出土时便有节，及凌云处尚虚心。"竹的精神，竹的风韵，竹的艺术，显示了中国光辉灿烂的文化，也显示了中华民族的情操和风采。我们的生活与竹子有着千丝万缕的关联，种竹、爱竹、恋竹、用竹是我们的共同夙愿，这种夙愿把我们的心紧紧地连在一起。早在80年代，我们便对浙江的大型竹编立屏《九龙壁》和《昭陵六骏》的精湛编技产生过浓厚的兴趣，对四川的瓷胎竹编产生过由衷的赞叹，对福建的镂空竹编产生过深切的关注，对湖南、广东的竹雕竹刻给予过高度的评价，认为这是我们祖国的艺术瑰宝，是中华民族的骄傲。近年来，中国的竹制艺术品有滑坡的趋向，身怀绝技的竹制艺人也出现了断层，一些80年代在竹子艺苑上产生过轰动效应的大型竹编精品也不再出现，这使我们感到叹惜、感到焦虑。

　　用照片和文字把这些珍贵的竹艺术奇葩定格下来，编一本精美的画册，向世人展示竹艺术的魅力和风采，不仅留存后世，而且使这一传统艺术得以继承、创新和发展都是很有意义的。1996年1月，一种共同的愿望使我们汇聚在南京林业大学竹材工程研究中心，开始了这项艰辛而有意义的工作。为尽量体现我国目前各地的竹工艺现状，我们携带摄影器械，拿着记事本子，足迹遍及钱江两岸，八闽大地，巴山蜀水，珠江南北，和各地的竹艺生产企业探讨，向竹刻老艺人求教，和编织设计人员研究……，用镜头摄下了他们巧手下的精品，用文字记下了他们心目中的思考。经拍摄和征集照片2000余张，从中精选出500余幅，按类整理，将竹编、竹刻、竹雕、竹材利用以及以竹为题材的绘画等内容归纳成竹篾天地、竹苑杂艺、竹材世界和翠竹清风四个部分，并配有系统而简要的文字叙述，对其精品、珍品还作了简介，力求全面展示中国竹工艺艺术的魅力和风采。

我们的拍摄、编著过程中，得到了多方面的热情关怀和支持。著名书法家、《人民日报》社社长邵华泽先生为这部画册题写了书名；中国林业科学研究院院长江泽慧教授为这部画册写了序；林业部、中国林业科学研究院、南京林业大学、浙江省林业厅、浙江省工艺品进口公司、浙江省安吉县林业局等单位为这部画册的出版给予了经济上的援助，各地的生产企业及诸多的先生们也为该画册的完成出了力。南京林业大学兼职教授，本画册副主编，日本神户捆包事业协同组合社长桂田 茂先生亦为画册的出版提供了有力的资助。书法家叶文祥为这部画册题写了篇名。还有余磊等24位先生及单位为我们提供了160余幅照片。没有他们的支持和帮助，这部画册就难以顺利地和大家见面，让我们以笔代躬，向他们致以诚挚的敬意和谢意。

本画册正文由王正、徐华铠撰写，由张新萍、周吉仲、乔海清负责中文翻译成英文的工作。

在画册付梓之际，我们更把深情的目光移向广大读者，希望这本画册能为大家了解和研究中国竹工艺艺术有所帮助和借鉴，也为弘扬中国这一传统艺术起到一些宣传推动作用，并诚恳地期待着大家的批评和指正。

《中国竹工艺》编委会
1996年10月

PREFACE

China is rich in bamboo resources which contain 400 odd species of 30 genera,and among the countries that have the most bamboo species and the largest yields in the world.The tall,straight and beautiful bamboo keep verdant all the year round,stand side by side and grow thickly on hills,valleys and river banks.They decorate our landscape with elegant green groves.

For thousands of years,bamboo has glittered in Chinese culture,arts and daily life.It is an important component of Chinese traditional culture.The bamboo handicraft articles embody the value of bamboo to the largest extent."When the technology is combined with bamboo,thousands of novel and fresh compositions appear."The Chinese bamboo artists,follow the nature and characters of bamboo itself,are creating series of bamboo handicraft by their intelligence and acumen.From exquisite and fine bamboo weaves to simple and plain bamboo decoration,from excellent bamboo body and joint handicrafts of superlative craftsmanship to bamboo root carving of uncanny workmanship,all of these reflect the bamboo's nature of elegance and plain, which give inspiration in the form of image and entertain people with aesthetic feeling.

In a poem, the bamboo was described as:"There are already joints before it grows out from the earth,and it keeps hollow even if it grows as high as touching the clouds".In Chinese,the words"joint"and"hollow"pronounce just like"moral integrity"and"modes"respectively.So the bamboo is often likened to good nature and characters. The spirit,graceful bearing and art of bamboo demonstrate China's magnificent culture and sentiment,and the elegant demeanor of the Chinese nation.We have a thousand and one links with bamboo.Planting bamboo, enjoying bamboo scenery and using bamboo material are the long-cherished wish of us.This wish closely links our hearts together.As early as 1980s,we took a great interest in the consummate weaving skill of large bamboo weaving screens"Nine Dragons Wall"and"Six Steeds of Zhao Tomb of Tang Dynasty"produced in Zhejiang Province.We gasped with admiration from the bottom of our hearts at Sichuan Province's bamboo weaving with porcelain roughcast.We paid close attention to Fujian Province's hollowed out bamboo weavings and set a high value on carved bamboo produced in Guangle County of Hunan Province.We consider them as gems of China and the pride of Chinese nation.In recent years,China's bamboo handicrafts tend to decline. Skilled bamboo handicraft artists have none to come after as successor.Some large bamboo weaving fine articles that made a furore in 1980s no longer appear.This makes us fell sigh and anxious.

It is of great significance to record these valuable works of bamboo art.The aim of this album is to show the charm and elegant demeanor of bamboo art,not only find its place in history,but also make this traditional

art carried forward,innovated and developed.In January of 1996, the common desire made us gathered in Bamboo Engineering Centre of Nanjing Forestry University and began this hard and significant work.In order to show the situation of bamboo technology all around China to the best,we took cameras and notebooks with us and walked everywhere, inquired into subjects with bamboo handicraft producers,learned from bamboo carving artists,studied with weaving designers,took pictures of fine handicrafts and recorded their ideas and intentions. We have taken and collected altogether more than 2000 pictures,some 500 from them have been selected and sorted according to the varieties of products,such as bamboo weaving,carving,engraving and articles of daily use,consequently four parts have been formed in this book,they are: the kingdom of bamboo strips,the variety of art and crafts in bamboo realm,the world of bamboo material,the gentle breeze over green bamboo.To all chosen masterpieces we attached brief and systematic notes,trying to demonstrate the essence and spirits of Chinese bamboo culture.

During the process of taking picture and compiling,we got enthusiastic supports and cares from various sides.Mr.Shao Huaze,a famous calligrapher and head of the People's Daily,calligraphed the title of this album, Professor Jiang Zehui,director of Chinese Academy of Forestry,wrote the foreword.Ministry of Forestry,Nanjing Forestry University,Zhejing Provincial Department of Forestry,Bamboo Center of Ministry of Forestry,Chinese Academy of Forestry,Arts and Crafts Import and Export Company of Zhejiang Province financially supported the publishing.Many enterprises and expertss made their contribution to this album.The guest professor of Nanjing Forestry University,deputy editor-in-chief of this album,president of Kobe Packing Cooperative Association, Mr.Shijiru Katsurada assisted us energetically.Without their support and help,this album would not come out smoothly.Here we would like to present our sincere compliments and heartfelt thanks to them.

All the photos are taken by Ma Youji,except those named otherwise.

On the occasion of publishing this album,we sincerely hope that this album can be used for reference in understanding and studying China's bamboo handicrafts arts,for enhancing Chinese traditional arts,Any candid criticism or suggestion is welcome.

Editorial Committee
CHINESE BAMBOO HANDICRAFTS
October,1996

中国竹工艺

目录 CONTENTS

序 (8)
FOREWORD

再版前言 (10)
PREFACE TO SECOND EDITION

前言 (12)
PREFACE

竹林风光 (19)
SCENERY IN BAMBOO GROVES

竹篾天地 (35)
A KINGDOM OF BAMBOO STRIPS

 竹篾技艺 (36)
 Bamboo splitting

 竹编图案集锦 (40)
 Collection of bamboo weaving patterns

 篮 (44)
 Baskets

 盘 (47)
 Plates

 瓶 (49)
 Vases

 茶具、酒具、烟具、文具、咖啡具 (57)
 Tea sets, drinking sets, smoking sets, stationery, coffee sets

 台屏、字、画与龚扇 (60)
 Table screens, calligraphy, paintings and fans

 灯具 (65)
 Lamps

 罐、筐、篓、包及其它 (66)
 Pots, baskets, boxes, etc.

 禽鸟 (68)
 Birds

 走兽 (76)
 Beasts

 灵禽神兽 (85)
 Legendary animals and birds

历史神话人物 (94)
Legendary and historical stories
建筑艺品(95)
Architecture

竹苑杂艺 (105)
THE VARIETY OF ART AND CRAFTS IN BAMBOO REALM

竹刻 (106)
Bamboo engraving
竹雕 (120)
Bamboo carving
民族竹用具 (136)
Bamboo tools of national minorities
竹制小工艺 (137)
Small bamboo handicrafts

竹材世界 (143)
THE WORLD OF BAMBOO MATERIAL

竹家具 (144)
Bamboo furniture
竹装潢 (152)
Bamboo decoration
竹建筑 (160)
Bamboo architecture

翠竹清风 (169)
GENTLE BREEZE OVER GREEN BAMBOOS

秦汉竹简 (170)
Bamboo letters of Qin Dynasty and Han Dynasty
历代名家部分画竹精品 (171)
Masterpieces of bamboo painting by famous artists through the ages

SCENERY IN BAMBOO GROVES

竹林風炎

云南省是中国乃至世界丛生竹的重点产区,盛产大型丛生竹。图为德宏傣族景颇族自治州低海拔河谷坝区的热带丛生竹林景观。
这是一个美丽的地方,巨竹成林环抱傣乡佤寨,翠影丛丛点缀彩云之南,人歌象舞吟咏边疆太平盛世,竹苞松茂吸引中外嘉宾云集。烧好竹筒饭,备上竹筒酒,吹响葫芦丝,跳起孔雀舞,"世界竹类故乡"展示着迷人的风采!(摄影:辉朝茂)
The landscape of big-size clustered bamboo which distributed in the tropical and subtropical flatland of lower elevation valley in Dehong, Yunnan province, China. (Photos: Hui Chaomao)

龙竹是东南亚地区栽培最广泛的大型丛生竹种,高度达15~25m,秆型高大,单位面积产量相当于毛竹的5~8倍。(摄影 辉朝茂)
Dendrocalamus giganteus, a big-size clustered bamboo, which is widely distributed in Southeast Asia. High is 15~25m of its culm. (Photos: Hui Chaomao)

刺破青天
Pointing to the sky

云南西南特产的珍稀竹种巨龙竹,是世界上最大的竹子,其秆径可达30厘米,高可达30米以上,堪称竹中极品。(摄影:杨宇明)
Dendrocalamus sinicus, the largest bamboo in the world, which is distributed in southwest Yunnan, China. Its culm can grow to more than 30 m tall with 30 cm diameter. (Photos: Yang Yuming)

巨龙竹秆形高大,犹如鹤立鸡群。(摄影:辉朝茂)
Dendrocalamus sinicus is standing head and shoulders above other bamboos. (Photos: Hui Chaomao)

生机勃发
Full of life

刚竹滴翠
Greenery drops from bamboo leaves

竹径通幽
A bamboo path leading to quietness and seclusion

云南地处祖国西南边陲，风景优美，气候宜人，竹类资源异常丰富，是世界竹类植物的起源地和现代分布中心之一，被誉为"世界竹类的故乡"。云南独特的珍稀竹种多样性、竹林景观多样性、天然群落多样性和民族竹文化多样性，越来越受到国内外的关注。图为分布在滇西高黎贡山国家自然保护区的珍稀竹种针麻竹，其紫红色大型头状花序为世界竹亚科所罕见。（摄影：辉朝茂）

The bamboo resources of Yunnan, which is situated in the southwest of China, is acknowledged as one of the original places and modern distribution center of bamboo plants, and praised "the home of bamboo of the world". There are 5 characteristics in bamboos diversity of Yunnan, as' well as valuable and rare germ plasm, floristic composition, natural bamboo forests, ecological landscape, bamboo culture of national minority. Figure is *Cephalostachyum scandens*, which is a rare and valuable species in the world. Its violet red florescence is seldom seen in bamboos all over the world. (Photos: Hui Chaomao)

藤本状竹类是滇西至滇西南特有的一种热带竹林景观，表现了丰富多彩的竹类多样性。（摄影：刘家柱）

The landscape of rattan–like bamboos, they are distributed in west to southwest of Yunnan. (Photos: Liu Jiazhu)

被列入中国重点保护植物的著名工艺竹种筇竹。（摄影：薛嘉榕）

Qiongzhuea tumidinoda, a famous bamboo species for handicrafts, which has been classified as the list of wild plants of national priority protection. (Photos: Xue Jiarong)

珍稀竹种香糯竹分布滇南地区，傣族群众用其幼秆烧制"竹筒饭"，清香可口，是竹类特殊利用方式之一。（摄影：辉朝茂）

Cephalostachyum pergracile, a rare and valuable bamboo species, which is distributed in south of Yunnan and used to cooking rice by Dai people. (Photos: Hui Chaomao)

黄金竹：黄金间碧玉，节间金黄色而嵌绿色条纹，观赏和工艺价值极高。

Bambusa vulgaris cv. VITTATA, a famous bamboo species for ornamental and handicrafts.

云南甜竹的鲜笋品质优秀，鲜嫩可口，具甜味，宜鲜食，有"甜笋"之称。

Dendrocalamus brandisii, a famous bamboo species for shoots with the best quality.

分布在滇南地区的苦笋竹的花序。
The inflorescence of *Indosasa amara*, which is distributed in south of Yunnan.

针麻竹的种子。
The seeds of *Cephalostachyum scandens*.

主产华南地区的优良经济竹种麻竹的花序。
The inflorescence of *Dendrocalamus latiflorus*, which is distributed in south of China.

珍稀竹种梨藤竹是滇西南地区特有竹种，其果实大如核桃。
The fruit of *Melocalamus arrectus*, which is distributed in southwest of Yunnan.

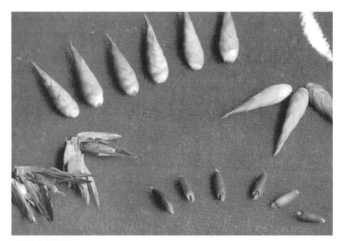

巨龙竹的种子。
The seeds of *Dendrocalamus sinicus*.

竹亚科植物开花周期较长，一般都要数十年才开花，有的百年不遇。图为分布在我国澜沧江下游地区的天然黄竹林开花的情景。
（摄影：辉朝茂）
Flowering landscape of nature bamboo forest of *Denrocalamus membranaceus*, which is distributed in the lower reaches of Langcanjiang river in Yunnan, China.
(Photos：Hui Chaomao)

我国滇西北横断山区是世界高山竹类的重点产区,大面积的高山竹林有效地保护着长江、澜沧江、怒江上游地区的生态环境。(摄影:安明超)
The Hengduanshan mountain area is the important distribution area of alp bamboo. The large area of alp bamboo forest protect the ecological environmental of the lower reaches of Yangtze River, Langcanjiang river, and Nujiang River efficiency. (Photos: An Mingchao)

植被破坏、洪水无情,但用竹子建成防护林体系,可有效地保护生态环境。(摄影:辉朝茂)
The bamboo forest can protect the ecological environmental efficiency. (Photos: Hui Chaomao)

天高地远、旷野苍茫,只有竹子抗击着大自然的风风雨雨,顽强地扎根在红土高原上。(摄影:辉朝茂)
Resisting the difficulties and hardships, bamboo growing between the land and sky indomitably. (Photos: Hui Chaomao)

THE KINGDOM OF BAMBOO STRIPS

竹箋天地

竹篾技艺 Bamboo splitting

卷节
Removing nodes

剖竹
Splitting culms

起间
Removing node walls

开间
Cleaning node walls

1. 竹篾加工

竹子坚实而富有弹性、韧性，劈裂性能好，很适宜劈篾编织。不同的竹种有不同的特性和用途，如毛竹、桂竹、淡竹、黄古竹、水竹、慈竹、单竹、青皮竹等，竹节平而疏，纤维坚韧，是优质篾用竹种。竹子经锯切、卷节、剖竹、开间、劈篾、刮篾、劈丝、抽丝及浑丝等多道工序能制成各种竹丝篾片。竹编就是竹丝篾片的挑压交织，一般称被挑压的篾为"经"，而编入的篾称为"纬"，由经与纬的挑压可编织出千变万化的图案，制作出千姿百态的竹制工艺品。

2. 竹编简史

中国的竹编历史悠久，据考古发现，早在新石器时代，人们就掌握了编织技术。浙江钱山漾良渚文化遗址中出土200件竹编器物，就足以佐证当时已有比较娴熟的编织技能。战国时期湖北楚墓出土的一把竹编扇子，扇面用细篾片编织成矩形图案，已相当精细。东晋时期浙江嵊县竹编就有"篾短秋翼蝉"的记载。明、清时期，竹编品种渐趋繁多，做工更为精致。竹编在民间已十分普遍。始创于元末明初的湖南益阳水竹凉席，盛行于明代的安徽舒城舒席已经成了当时的特产。清代同治年间，四川成都地区将竹编和瓷器、漆器相结合，创出了瓷胎竹编新工艺。艺人们用细如发丝的竹篾，在酒具、烟具、茶具、花瓶等器具上进行编织，典雅别致，成为四川竹编的一大特色。目前除成都外，渠县、自贡等地也有新的发展。浙江的东阳竹编在清代之前就选为贡品。民国初年，东阳艺人马富进编织的"魁星像"，在1929年西湖博览会上受到高度评价。清末民初，竹编工艺品还远销欧美国家。建国以来，嵊县（现为嵊州市）竹编发展更为迅猛，在动物造型及竹材的新工艺上均出现了欣欣向荣的局面。

中国历代的竹编艺人们通过对竹丝篾片的挑、压、弹、插、绕、穿、贴等技法，创造出数以百计的竹编技法图案，常见的有十字编、人字编、圆面编、装饰编及弹插编织等。竹编艺品的色彩也较丰富，有的表现淡雅清新、有的表现富丽庄重，一般以体现竹材本身的质地为上乘。

3. 竹编器具类

竹编工艺目前有器具类、欣赏类及屏风、壁挂等三大类别。

器具类包括篮、盘、瓶、罐、箱、盒、席、帘、扇等。

传统的民间工艺竹篮大多呈椭圆、长方、八角形，编织相当精细，有的在每3厘米长度内可排列120根篾丝，由于层层相套，又称"塔篮"或"脱篮"。许多精品往往将编织、雕刻、绘画及造型融为一体。20世纪80年代以来，为了旅游事业发展的需要，艺人们巧妙地把动物形态和实用性的竹篮结合为一体，编制出鸡篮、鸭蓝、鹅蓝等许多动物形竹篮，生动、美观、实用，惹人喜爱。

竹编盘子花样繁多，能编织成多种造型和图案，如扇盘、叶盘、象盘、鸭盘、鹅盘、鸡盘等，不胜枚举。福建、浙江有一种竹编篮胎漆盘，是用涂漆工艺制成，亮度高、漆膜厚、耐水泡，宛如瓷器。又比瓷器牢固轻巧，是馈赠亲友的极好佳品。

竹编花瓶起源于20世纪50年代，是一种新的竹编造型艺术。竹编花瓶精巧美观，多姿多态，尤以浙江、福建、四川最具特色。浙江花瓶编织品种繁多，其中以穿丝编最为精美。近年来，艺人们创造了"花筋工艺"，即在插筋的篾片上套印上各种图案花纹，使花瓶陡增生气，具有装饰美。后在花筋工艺基础上又发展"烫金工艺"，使竹编花瓶更加富丽堂皇。福建竹编花瓶多用弹篾编织，有人字弹、菠萝弹、横弹、孔雀尾弹，具有浮雕图案效果。还有一种雀目编，可以透过几何状编织空隙，看到内部的瓷瓶；四川成都地区的竹编花瓶，大多以瓷为胎，篾丝紧扣瓷胎，所有接头都藏而不露，浑然一体，并在花瓶面上编出如"双龙戏凤""熊猫""福、禄、寿、喜"等多种字画图案，精巧绝伦。四川渠县的薄胎细竹编更让人爱不释手，可谓一绝。被国际友人赞为"东方艺术之花"。

其他如箱、罐、盒等竹编艺品，也造型各异，各尽其美。

竹席常用毛竹为原料，精致者多用水竹。湖南益阳的凉席和安徽舒城的舒席用料采用水竹加工编织而成。篾纹纤细，平软光滑，色泽晶莹透明，散热性强。其编织异常精细，可在3厘米长度内排列10~30根篾丝，通过篾片的挑压还可编织出花鸟虫鱼、山水书画等的图案。益阳水竹凉席19世纪20年代就远销国外。安徽舒席近百年来也多次在国际展览中获奖。中国竹席也在不断创新，近年来出现了用竹片制成方块状，用丝或尼龙绳串联，称为"麻将"竹席，凉爽宜人。

竹帘用于门帘、窗帘，始于宋代，原产四川梁平，后发展为室内工艺品。精致者，每尺竹帘坯用竹丝可达1000根左右。目前品种用"四条屏""屏风""通景屏""帐檐""对联"等百余种。四川、浙江、福建等地所产竹帘还绘或编有山水、花鸟、人物等，更是锦上添花。

竹编扇子品种很多，有粗细之分，粗者称"篾把扇"，由粗篾编织而成，细者称"竹丝扇"，四川自贡的"竹丝扇"最为名贵，人称"龚扇"，是清末艺人龚爵五的绝技，现已后继有人，并出现了国家级工艺美术大师龚玉文。"龚扇"薄如绢纱，明亮透光，并编有人物、山水、花鸟、书法等图案。它以慈竹为原料，将竹篾刮得细如发丝，在半径12厘米长的团扇上，密排700多根竹丝作经线，又以同样细的竹丝作纬线，一根纬线需穿插700多次，所成扇面如丝绸一般。苏杭一带的折扇，其扇骨是用竹材做的，可折叠，品种繁多，精致华贵。

4. 竹编欣赏类

欣赏类竹编有动物、人物、仿古建筑等。竹编动物于20世纪60年代问世，是由浙江嵊州的中国工艺美术大师俞樟根在有关人员的配合下创造的，从此开辟了竹编动物的新天地。编织动物先行造型设计，依照动物的胎模（木模模型或用纱帐布制胎模）编织成形。艺人们用增减篾丝的方法把动物肌肤的凸凹面充分的编织出来。如嵊州竹编"长颈鹿"、东阳竹编"牛斗虎"、新昌竹编"斗牛"，将动物的肌肉表现得淋漓尽致。东阳竹编大象长4.3米，高2.16米，可谓庞然大物。竹编禽类，多用长短篾片仿照禽类的硬羽毛进行弹插，形象逼真。嵊州、新昌竹编的飞禽受到世人称赞，嵊州竹编"白尾海雕"1979年进入美国白宫。

为了增添动物的色彩，艺人们将漂白、花筋等新工艺运用于竹编动物。"漂白"即将竹篾进行脱脂漂白，竹编"白孔雀""仙鹤"即用漂白篾片编织而成，很有表现力。

竹编人物题材多样，有历史神话故事、历史英雄人物、现代人物等。编织粗细并茂、形神兼具。如东阳竹编"渔翁"、嵊州竹编"岳飞"、新昌竹编"老寿星"等均为竹编中的杰作。

竹编仿古建筑也极精致传神，浙江竹编"六和塔""天坛""花塔"等被国际友人评为"东方珍宝""世上精品"。

5. 竹编屏风类

竹编屏风类，有立屏和挂屏、折叠屏之分。浙江嵊州竹编《昭陵六骏》和东阳竹编《九龙壁》大型立屏堪称绝品。《昭陵六骏》取材于中国唐代皇帝李世民昭陵墓前的石刻《昭陵六骏》。作品在长8.22米，高2.35米，底座宽0.68米的铜绿色编织立面上，六匹骏马以高浮雕的形态，运用100多种技法编制而成，是我国目前最大的立屏。大型立屏《九龙壁》是从中国古建筑"影壁"

劈篾
Striping

劈丝
Making threads

用嘴配合劈丝
Making threads

抽篾
Processing strips

刮篾
Processing strips

编织
Weaving

串丝
Arranging threads

粗编织
Rough weaving

精细编织
Fine weaving

装配
Assembling

"照壁"与中国传统落地插屏中脱颖而出。作品长 6.19 米，高 2.68 米，底座宽 0.55 米，由底座、束腰、显口、边框、堂板、腰斗、扑斗、压栋等八大部分组成。构图中心为九条龙、八个火球、五十一片云朵及连绵起伏的波涛围绕其中，作品采用 150 余种编织手法，用每厘米 50 余根篾丝编织，八个火球用八种编织图案，对九种不同造型的龙，用九种不同编织图案，龙头、龙发、龙尾，用弹、插、胶、卷、雕等手法完成。整个主体造型多变、转折多，很多技法是竹编史上的新水平。该精品获得 1984 年全国工艺美术金杯奖，并荣获国家珍品称号。

6. 各地竹编特色

中国竹编工艺在长期的发展过程中，从日用品发展到工艺美术品，表现了艺术巨匠们的创新精神和高度智慧，同时也形成了许多地方特色。大而观之，浙江竹编以精巧细腻，造型美观，高精产品迭起，花色品种繁多而著称，目前竹编品种多达 10000 余种。嵊州、东阳、新昌、浦江、乐清、武义、鄞县、杭州等地为编产地，尤以嵊州、东阳、新昌竹编最为出色，嵊州有"中外竹编第一家"的称誉。四川是中国重点竹编产地，竹编不仅精细，而且能充分表现竹材的自然美。以瓷胎竹编和竹丝扇最为著名。成都、崇州、自贡、渠县、江安、梁平、丰都、安岳等地都是竹编产地。福建竹编是中国竹编出口大省，产品多以毛竹为原料，粗细兼备，并用弹簧点缀，竹篾色泽深沉，主要产地有泉州、莆田、厦门、福州、漳州、安溪、宁德、闽侯、古田等。云南竹编具有浓厚的少数民族风情，多饰以民族特色的花纹，竹编以日用品为主。江苏、安徽、广东、广西、湖南、湖北、江西、贵州、台湾等地的竹编也形成自己的特色。目前，中国竹编花色品种已发展到数万种，各种流派纷呈多姿，争奇斗妍，真正成了一个竹篾编织的绚烂天地。

1. Bamboo strip Processing

The culm of bamboo is hard, tough and flexible, but it is easy to be cut into strips for weaving. Different bamboo species, such as *Phyllostachys pubscens*, *Ph. angusta*, *Ph. heteroclada*, *Sinocalamus affinis*, *Lingnmania chungii*, *Bambusa lextills* can be used for making different woven articles. These species have very flat nodes and great internodal lenth, with fine and strong tissue. Through cutting, node removing, sectoring, splitting and other operations bamboo stripss and threads are produced. The strips and threads are basic material for weaving various bamboo crafts.

2. A Brief History of Bamboo Strip Weaving

The history of bamboo weaving in China can be traced to the Neolithic Age, in the Yanliangchu Ruin of Zhejiang Province more than 200 bamboo woven articles have been unearthed, which demonstrated comparatively skilled weaving techniques. A bamboo fan, unearthed from Chu Ruin of the Warring States Period (403~221 B.C.), of Hubei Province, was woven of thin strips into rectangle pattern, showing fine workmanship. The bamboo woven products in Shen County of Zhejiang Province enjoyed a high reputation early in East Jin Dynasty (317~420 A.D.) which were famous for their thin strips as transparent as a cicada's wing. Bamboo woven products diversified step by step during Ming (1368~1644 A.D.) and Qing (1636~1911 A.D.) dynasties, with improved techniques, they became quite popular among ordinary people. The bamboo mats made of *Ph. angusta* in Yiyang County. Hunan Province at the end of Yuan Dynasty (1271~1368 A.D) and the beginning of Ming Dynasty (1368~1644 A.D.), the Shu mats made in Shucheng County, Anhui Province were regarded as special products. During Tongzhi reign of Qing Dynasty the masters in Chengdu Prefecture of Sichuan Province combined bamboo weaving with porcelain and lacquer ware, created bamboo woven articles with a porcelain body. Using threads as thin as hair workers wove them on drinking vessels, tea sets and vase, and made bamboo ware of Sichuan original style, along with Chengdu Prefecture, people in Quxian and Zigong counties are also developing such kind of products. Bamboo weaving in Dongyang County. Zhejiang province were selected as tributes offered to emperors even before Qing Dynasty. A bamboo statue God of Literature made by a famous master Ma Fujin in Dongyang County was valued highly on the West Lake Exhibition in 1929. Bamboo woven products were exported to Europe and America at the end of Qing Dynasty and the beginning of the Republic of China. The bamboo woven products in Shengxian County have developed rapidly since the foundation of the People's Republic of China, the shaping of animal figures and the new method of bamboo treatment have made great progress.

Generations of masters of bamboo weaving, by means of stitching, layering, inserting, winding stringing and pining, have created hundreds of pattern, most popular of them are cross weaving, v - weaving, circular weaving. The color of woven articles are different, some are dim and elegant, others are rich and brilliant, but the must preferred are rich and brilliant, but the most preferred are rich and brilliant, but the most preferred and those, which maintain the natural grain.

3. Bamboo Woven Tools

The bamboo woven articles are made for both daily use and decoration. The woven articles of daily use are baskets, plates, trays, dishes, bottles, jars, boxes, mats, curtains, fans.

The traditional bamboo baskets are of elliptic, rectangle and octagonal shape, of very fine workmanship, sometimes as many as 120 thin threads can be woven in a width of 3 mm. The are also basket sets which consist of several baskets of the same shape but different size, placed one into another. In some of the woven articles the art of weaving, sculpture and painting are combined. Since 1980 new types of bird shaped bamboo woven baskets have been created as a result of the development of tourist industry, they are cock basket, duck basket, goose basket and others, they are elegant, attractive and practical.

Bamboo woven plates, trays and dishes have diversified rapidly and effectively, there are many models and patterns, such as fan plate, leaf plate, elephant plate, duck dish, goose dish, cock dish. A type of bamboo woven baskets covered with lacquer are develpoed in Fujian and Zhejiang provinces, they are luminous and waterproof, bright like porcelain but not so heave, they sold well as tourist souvenir.

Bamboo woven vases were first developed in 1950s by means of a new modeling technology. Various bamboo vases are colorful and beautiful, those made in Zhejiang, Fujian and Sichuan provinces are most attractive. Many kinds of bamboo vases are produced in Zhejiang, a "bamboo printing technology" has been invented to cover vases with original patters and grains, and a "gold stamping technology" with more rich surface. The specific weaving techniques give vases a relief surface, and sometimes the porcelain body can be seen through bamboo weaving. The vases produced in Sichuan are famous for their original surface design, such as " Two dragons playing with a phoenix", "Panda", "Blessedness, Richness, Lonegvity, Happiness". The material for weaving is treated properly, which does not suffer from wetness, dryness, deformation and insect attack. Bamboo woven vases are praised by foreigners as "a flower of oriental art".

Bamboo woven jars, boxes and cases are also very beautiful and practical.

Bamboo mats are generally made of *Ph. Pubescens*, but the finer ones are of *Ph. heteroclada*. The mats made in Yiyang County of Hunan and Shucheng County of Anhui are made of *Ph. heteroclada*, they are very smooth, brilliant and cool, as a result of fine workmanship, ten to thirty threads can be arranged in a width of 3 mm. Through various weaving methods many designs can be expressed on mats, such as flowers, birds, fish, mountains and rivers. The mats of *Ph. heteroclada* from Yiyang exported as early as in 1920s, the Shu mats from Shucheng County have been awarded for several times on international exhibitions. A new type of bamboo mats made of small rectangular pieces of bamboo, piled together with silk or nylon threads, these mast are called "Domino".

Bamboo curtains are hanged to cover doors or windows, the production of bamboo curtain, originated in Liangping County of Sichuan, began in Song Dynasty. The weaving techniques reached a very high level, within a width of 30 cm masters arrange some 1000 fine bamboo threads. More than one hundred kinds of bamboo curtains are being produced. The curtains made in Sichuan, Zhejing and Fujian provinces are painted with landscapes, flowers, birds, and human figures.

There are two kinds of bamboo woven fans, rougher and thinner, the rougher ones are made of bamboo strips, while the thinner ones of bamboo threads. Bamboo thread fans produced in Zigong County of Sichuan are most famous, called "Gong Fan", which were firstly developed by a master named Gong juewu at the end of Qing Dynasty. Gong fans are as thin as sheer silk, and transparent with different patterns like human figures, landscapes, flowers, birds, calligraphy. Gong fans are made of *Sinocalamus affinis*, the threads are hairy thin, more than 700 wefts are arranged in width of 12 mm. The surface of woven fans is thin and brilliant like silk. Folding fans in Suzhou and Hangzhou with bamboo backbone are also very famous and expensive.

4. Bamboo Woven Articles for Decoration

The bamboo weaving for decoration includes animlas, figures, imitations of ancient building. Bamboo-woven animals began to come out in 1960s, which were created by Yu zhanggen, an arts and craffs master from Shengzhou, Zhejing Province, in cooperation with other artists, thus, a new field of bamboo arts were opened. Before weaving, it is needed to design the shape of animals, then the shapes are woven in accordance with animal's roughcast models (made of wood or gauze). The artists weave the skins of animal concavely or convexly by means of increasing or decreasing the bamboo strips. Shengzhou's bamboo weaving "Giraffe", Donyang's "Bull fighting Tiger" and Xinchang's "Bull Fighting" portray the muscle of animal incisively and vividly.

上漆
Lacquering

瓷胎编织
Weaving upon a porcelain body

Dongyang's bamboo-woven elephant is 4.3m in length and 2.16m in height. It is really a huge creation. For bamboo-woven birds, uneven bamboo strips are used to simulate the hard plumage of bird. Shengzhou and Xinchang's bamboo-woven birds were highly praised. Shengzhou's bamboo-weaving "White-tailed Eagle" was collected by White House of U.S.A.

In order to make the animals colorful, the artists utilize the new technology of bleaching and strip-coclornig.

Bleaching means to make the bamboo strips degreased and bleached. The bamboo weavings "White peacoch" and "Crane" are woven with bleached bamboo strips, which have good manifestation.

The theme of bamboo-woven figure is varied, including historic fairly tales, historic heroes, modern figures, etc..

Some of them are succinct and the others are exquisite. All of them are excellent both in appearance and spirit. Dongyang's bamboo weaving "Fisherman", Shengzhou's "Yue Fei" (a hero in Song Dynasty) and Xinchang's " God of Longevity" are all the masterpieces of bamboo weaving.

A masterpiece of bamboo woven article "a pearl returning to the sea " was presented as a souvenir by Zhejiang Provincial Government to Macao Special Administrative Region Government in 1999.

Bamboo-woven imitations of ancient building are also exquisite and vivid. Zhejiang's bamboo weavings "Liuhe Tower"," Temple of Heaven" and "Tower of Flowers" are appraised as "Oriental Treasure" and "Masterpiece of Art".

5. Bamboo Woven Screens

Bamboo-woven screens can be divided into floor screen, hanging screen and folding screen. Shengzhou's large bamboo weaving screen "Six Steeds of Zhao Tomb of Tang Dynasty" and Dongyang's "Nine Dragons Wall" may be rated as masterpieces through the ages. The image of "Six Steeds of Zhao Tomb of Tang Dynasty" is drawn from the carved stone "Six Steeds of Zhao Tomb" in front of Emperor Li Shimin's Tomb of Tang Dynasty. Six steeds were woven in form of high relief by more than 100 techniques of artistic expression on a large bamboo weaving base which size is 8.22 m in length, 2.35 m in height and the width of its base is 0.68 m. This is the largest floor screen in China up to date. This large floor screen "Nine Dragons Wall" consulted the wall screen in Chinese ancient structure and traditional floor table plaque and come to the fore. This works of art is 6.19 m in length, 2.68 m in height and the width of its base is 0.55 m, and composed of eight parts. The main body of the composition is 9 dragons, with 8 fire balls. rolling waves and 51 pieces of clouds around them. This works of art adopted more than 150 weaving techniques. The bamboo strips used are as thin as 0.2 mm in width. The 8 fire ball are woven with 8 different models, 9 weaving patterns are used. The heads, hairs and tails of dargons are woven with the techniques of flicking, inserting, gluing, rolling and carving. The modeling of main bady is varied. Many techniques are creative in the history of bamboo weaving. This masterpiece won the gold medal of national arts and handicrafts in 1984 and the title of "national treasure".

6. Characteristics of Bamboo Weavings from Different Areas

In the process of its development. China's bamboo weaving technique has expanded from articles of daily use to arts and handicrafts articles. These works of art demonstrate the consciousness of bringing forth new ideas and the great intelligence of artists. Simultaneously, the local characteristics were formed. Generally speaking, Zhejiang's bamboo weavings are known for their exquisteness, their beautiful modeling, the quantity and variety of fine articles. At present, the quantity of bamboo weaving is as many as over 10 thousands. Shengzhou, Dongyang, Xinchang, Pujiang, Leqing, Wuyi, Yinxian, Hangzhou counties are all the places of producing bamboo weaving, in which Shengzhou, Dongyang and Xinzhou's bamboo weavings are the best. Shengzhou have enjoyed the reputation of being"No. One of bamboo weaving". Sichuan is a key province of bamboo weaving, its bamboo weaving show not only extreme exquisites, but also the natural beauty of bamboo. The most famous articles of bamboo weaving in Sichuan are bamboo weaving with porcelain roughcast and thin bamboo-woven fan. Chengdu, Chongqing, Zigong, Quxian, Jiang'an, Liangping, Fengdu, Anyue counties are also the places of bamboo weaving. Fujian Province is the key province wihch produces exported bamboo weaving. The raw material is *Phyllostachys pubescens* mainly, which is varied in thickness, and decorated with ficking strips. The bamboo strips are characterized for their dark color. The main places of producing bamboo weaving in Fujian Province are Quanzhou, Putian, Xiamen, Fuzhou, Zhangzhou, Anxi, Ningde, Minhou and Gutian counties. Yunnan's bamboo weavings have the strong flavor of minority nationality. They are often decorated with figures of nationality characteristics. The bamboo weavings there are mainly articles of daily use. The bamboo weavings in Jiangsu, Anhui, Guangdong, Guangxi, Hunan, Hubei, Jiangxi, Guizhou and Taiwan provinces have the characteristics of themselves. Now the variety of China's bamboo weaving numbers tens of thousands. Various schools of technique blossom radiant splendor and form a splendid kingdom of bamboo weaving.

经纬交织 气象万千
——竹编图案集锦 Collection of bamboo weaving patterns

竹丝篾片坚实而富有韧性，很适宜编织。一般被挑压的篾称为"经"，编入的篾称为"纬"，经与纬的交织，可编出千姿百态的图案，这里撷取的是常见的竹编图案。

Bamboo strips are substantial and tough, which are suitable for weaving. Strips picked in operation are regarded as warps, while those put statistically are wefts. The combination of warps and wefts creates varied patterns. The following ones are most popular.

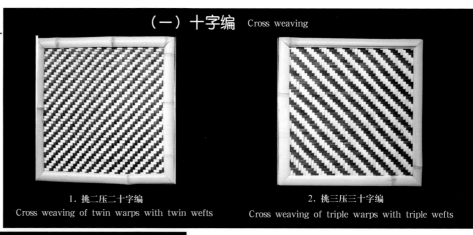

（一）十字编 Cross weaving

1. 挑二压二十字编
Cross weaving of twin warps with twin wefts

2. 挑三压三十字编
Cross weaving of triple warps with triple wefts

3. 单篾十字编
Cross weaving of single strips

4. 穿篾菱花块
Weaving of rhombus pattern with jointed strips

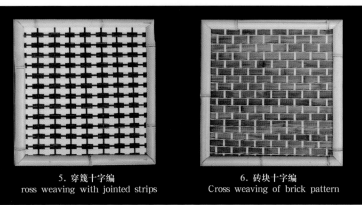

5. 穿篾十字编
Cross weaving with jointed strips

6. 砖块十字编
Cross weaving of brick pattern

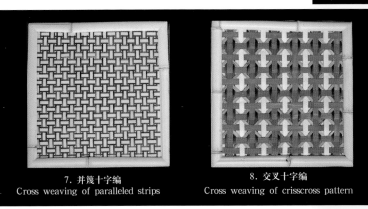

7. 并篾十字编
Cross weaving of paralleled strips

8. 交叉十字编
Cross weaving of crisscross pattern

（二）六角编 Hexagon weaving

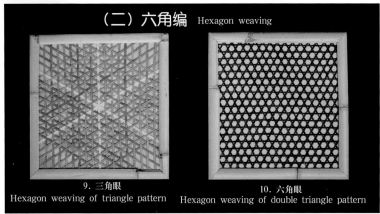

9. 三角眼
Hexagon weaving of triangle pattern

10. 六角眼
Hexagon weaving of double triangle pattern

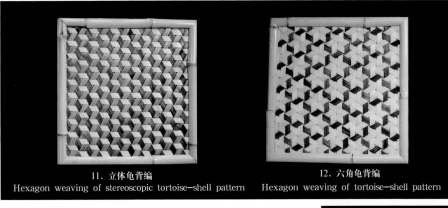

11. 立体龟背编
Hexagon weaving of stereoscopic tortoise-shell pattern

12. 六角龟背编
Hexagon weaving of tortoise-shell pattern

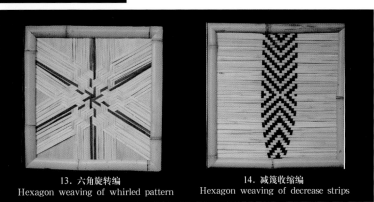

13. 六角旋转编
Hexagon weaving of whirled pattern

14. 减篾收缩编
Hexagon weaving of decrease strips

（三）图案花编 Weaving designs

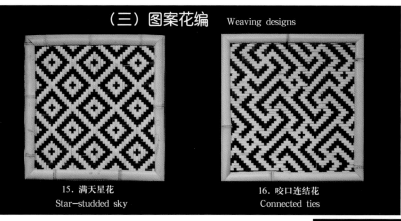

15. 满天星花
Star-studded sky

16. 咬口连结花
Connected ties

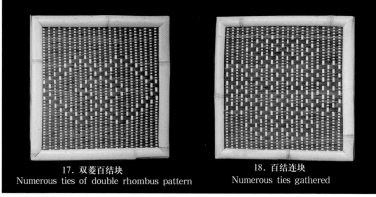

17. 双菱百结块
Numerous ties of double rhombus pattern

18. 百结连块
Numerous ties gathered

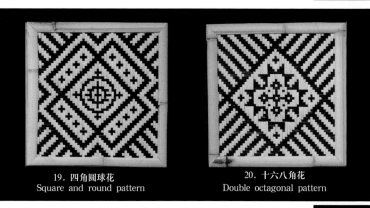

19. 四角圆球花
Square and round pattern

20. 十六八角花
Double octagonal pattern

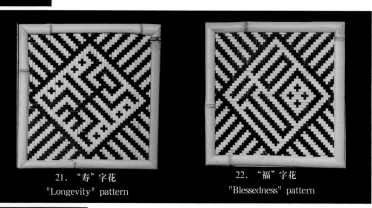

21. "寿"字花
"Longevity" pattern

22. "福"字花
"Blessedness" pattern

（四）穿篾穿丝编 Weaving of strips and threads

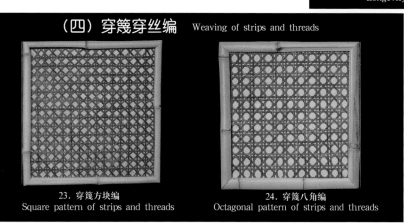

23. 穿篾方块编
Square pattern of strips and threads

24. 穿篾八角编
Octagonal pattern of strips and threads

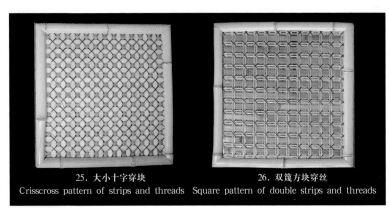

25. 大小十字穿块
Crisscross pattern of strips and threads

26. 双篦方块穿丝
Square pattern of double strips and threads

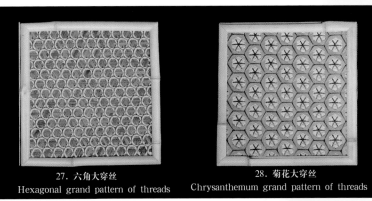

27. 六角大穿丝
Hexagonal grand pattern of threads

28. 菊花大穿丝
Chrysanthemum grand pattern of threads

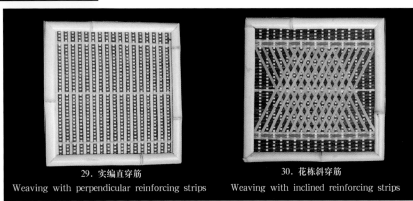

29. 实编直穿筋
Weaving with perpendicular reinforcing strips

30. 花栋斜穿筋
Weaving with inclined reinforcing strips

（五）螺旋圆形编 Spiral round pattern

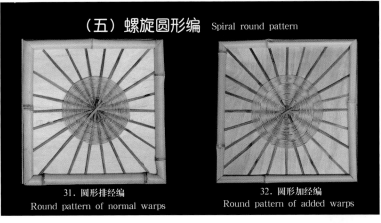

31. 圆形排经编
Round pattern of normal warps

32. 圆形加经编
Round pattern of added warps

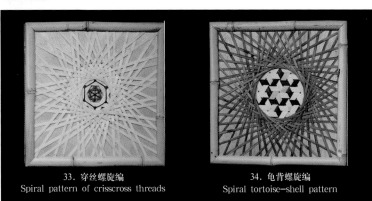

33. 穿丝螺旋编
Spiral pattern of crisscross threads

34. 龟背螺旋编
Spiral tortoise-shell pattern

（六）弹花编 Catapult pattern

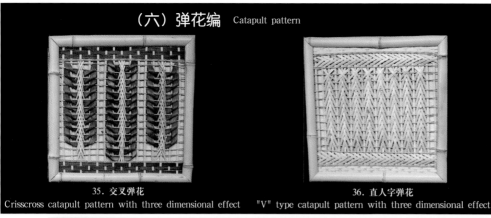

35. 交叉弹花
Crisscross catapult pattern with three dimensional effect

36. 直人字弹花
"V" type catapult pattern with three dimensional effect

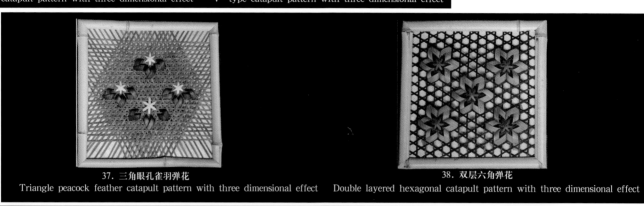

37. 三角眼孔雀羽弹花
Triangle peacock feather catapult pattern with three dimensional effect

38. 双层六角弹花
Double layered hexagonal catapult pattern with three dimensional effect

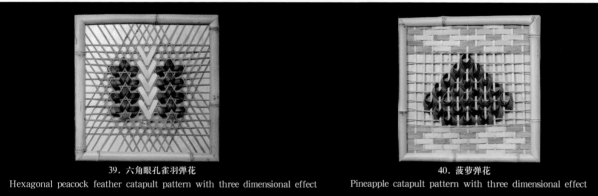

39. 六角眼孔雀羽弹花
Hexagonal peacock feather catapult pattern with three dimensional effect

40. 菠萝弹花
Pineapple catapult pattern with three dimensional effect

民间工艺篮（浙江嵊州）Baskets of folk art (Shengzhou, Zhejiang)
民间工艺竹篮在明清时期流传于浙江一带，编织精细、费时。大多层层相套，形如宝塔，故又称"塔篮"、"脱篮"或"托篮"。这类竹篮不仅可以装食品，放花果，而且可存衣、藏书。常见的品种有：套篮、食篮、香篮、考篮、花篮、挈篮、鞋篮和烟篮。
Bamboo baskets of folk art were quite popular in Zhejiang area in Ming and Qing dynasties. The are made through fine and laborious skills, designed to be put on one another, they look like a pagoda, therefore, are called "pagoda baskets". These baskets hold not only food, but also others. There are food baskets, incense baskets, book baskets, flower baskets and fruit baskets and many others.

串丝花篮（浙江新昌）Flower baskets of stringed threads (Xinchang, Zhejiang)

民间工艺篮(浙江嵊州)
Baskets of folk art (Chengzhou, Zhejiang)

民间工艺篮(浙江嵊州)
Baskets of folk art (Chengzhou, Zhejiang)

串丝花篮与果盘(浙江新昌) Flower baskets of stringed threads and a fruit tray (Xinchang, Zhejiang)

"母子大花篮"(浙江东阳) Big flower baskets (Dongyang, Zhejiang)

小方篮（福建泉州）
A small square basket (Quanzhou, Fujian)

小提篮（福建漳州）
Small hand-baskets (Zhangzhou, Fujian)

花篮与八角果盘（浙江新昌）
A flower basket and octagonal fruit trays (Xinchang, Zhejiang)

小方篮与鞋形篓（福建漳州） A small square basket and shoe-shaped baskets (Zhangzhou, Fujian)

小花篮（浙江东阳）
Small flower baskets (Dongyang, Zhejiang)

小花篮（福建漳州）
Small flower baskets (Zhangzhou, Fujian)

串丝花篮（浙江新昌） Flower baskets of stringed threads (Xinchang, Zhejiang)

竹编提篮（浙江东阳）
Woven hand-baskets (Dongyang, Zhejiang)

串丝民间工艺篮（浙江新昌） Baskets of folk art of stringed threads (Xinchang, Zhejiang)

漆盘（浙江新昌）
Lacquered plates (Xinchang, Zhejiang)

串丝盘与盒（浙江新昌） Plates of stringed threads and a boes (Xinchang, Zhejiang)

串丝盘（浙江新昌）
Plates of stringed threads (Xinchang, Zhejiang)

盘 Plates

串丝盘与盒（浙江新昌）
Plates of stringed threads and a box (Xinchang, Zhejiang)

果盘(浙江东阳)
Fruit trays (Dongyang, Zhejiang)

扇形盘(浙江嵊州)
Fan-shaped plates (Chengzhou, Zhejiang)

桃形盘、叶形盘与菠萝盘(浙江嵊州)
A peach-shaped plate, a leaf-shaped plate and pineapple plates (Chengzhou, Zhejiang)

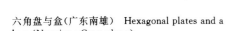

串丝盘(浙江新昌)
Plates of stringed threads (Xinchang, Zhejiang)

六角盘与盒(广东南雄) Hexagonal plates and a box (Nanxiong Guangdong)

龙纹瓷胎花瓶（四川成都） Porcelain-bodied vase with dragon patterns (Chengdu, Sichuan)

龙纹瓷胎大花瓶（四川成都） A big porcelain-bodied vase with dragon pattern (Chengdu, Sichuan)

花仙瓷胎大花瓶（四川成都） A porcelain-bodied vase with flower and angel pattern (Chengdu, Sichuan)

艺人们用四川盛产的慈竹，经严格加工后，在花瓶的瓷胎上进行精编，编织时篾丝紧扣瓷胎，依胎成型。艺人们在密编的基础上又创造出龙纹图案，显得更为珍贵。
Well-processed bamboo which grows in Sichuan Province was used to weave on the porcelain roughcast of vase. The bamboo strips nestle closely to the roughcast and copy the shape of roughcast. On the basis of close weaving, the artists created the dragon figures which makes this work of art more valuable.

瓷胎花瓶（四川成都） Porcelain-bodied vases (Chengdu, Sichuan)

瓷胎花瓶（四川成都）
Porcelain-bodied vases (Chengdu, Sichuan)

龙纹瓷胎大花瓶（四川成都） A big porcelain-bodied vase with dragon pattern (Chengdu, Sichuan)

瓷胎小号花瓶（四川成都） Small porcelain-bodied vases (Chengdu, Sichuan)

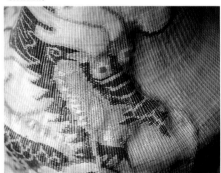

龙纹瓷胎大花瓶局部（四川成都） The detail of A big porcelain-bodied vase with dragon pattern (Chengdu, Sichuan)

瓷胎中号花瓶（四川成都） Porcelain-bodied vases of middle size (Chengdu, Sichuan)

瓷胎大花瓶（四川成都）
Big porcelain-bodied vases (Chengdu, Sichuan)

瓷胎花瓶（四川成都）
Porcelain-bodied vases (Chengdu, Sichuan)

瓷薄胎双龙花瓶（四川渠县）
A thinly porcelain-bodied vase with double dragon pattern (Quxian, Sichuan)

瓷胎花瓶（四川渠县）
Porcelain-bodied vases (Quxian, Sichuan)

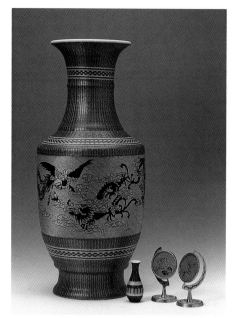

瓷胎龙凤花瓶（四川渠县） A porcelain-bodied vase with dragon and phoenix pattern (Quxian, Sichuan)

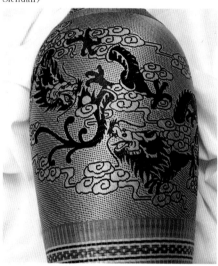

瓷薄胎双龙花瓶局部（四川渠县） The detail of a thinly porcelain-bodied vase with double dragon pattern (Quxian, Sichuan)

瓷薄胎熊猫花瓶（四川渠县） Thinly porcelain-bodied vases with panda pattern (Quxian, Sichuan)

瓷薄胎熊猫花瓶局部（四川渠县）
The detail of a thinly porcelain-bodied vase with panda pattern (Quxian, Sichuan)

弹篾花瓶（福建泉州）
Vases of suppressed strips (Quanzhou, Fujian)

弹篾花瓶（福建泉州）
Vases of suppressed strips (Quanzhou, Fujian)

双头孔雀花瓶（福建泉州） A vase with peacock pattern (Quanzhou, Zhejiang)

"弹篾花瓶"、花篮（福建泉州） A vase of suppressed strips and a flower baskets (Quanzhou, Fujian)

弹篾花瓶（福建泉州）
Vases of suppressed strips (Quanzhou, Fujian)

弹篾渴龙双耳花瓶(福建泉州)
A vase of suppressed strips with two thirsty dragons as ears (Quanzhou, Fujian)

弹篾渴龙双耳花瓶局部(福建泉州) The detail of a vase of suppressed strips with two thirsty dragons as ears (Quanzhou, Fujian)

弹篾渴龙双耳花瓶局部(福建泉州) The detail of a vase of suppressed strips with two thirsty dragons as ears (Quanzhou, Fujian)

细细的瓶颈两边,两条饮水的渴龙直立两边,得体地成了花瓶的耳环,瓶体的编织采用弹花,浑厚豪放,和精细的编织相映成趣。
Two thirsty dragons are attached to both sides of the slender neck, as if willing to drink water, forming suitable ears of the vase. The bold modeling and the fine weaving contrast pleasingly with each other.

龙耳花瓶(福建泉州)
A vase with dragon ears (Quanzhou, Fujian)

大肚花瓶(福建泉州)
A vase with a narrow neck and a big capacity (Quanzhou, Fujian)

弹篾花瓶（福建泉州）
Vases of suppressed strips (Quanzhou, Fujian)

竹贴花瓶（福建泉州）
Bamboo veneered vases (Quanzhou, Fujian)

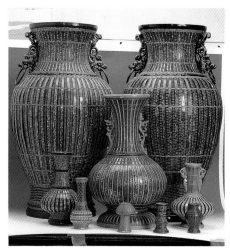

插筋花瓶（浙江嵊州）
Vases (Chengzhou, Zhejiang)

葫芦大花瓶（福建泉州）
A gourd-shaped vase (Quanzhou, Fujian)

葫芦大花瓶局部（福建泉州） The detail of a gourd-shaped vase (Quanzhou, Fujian)

双环花瓶（福建泉州）
A double-ringed vase (Quanzhou, Fujian)

蟠龙花瓶(福建泉州)　Vases with a coiling dragon (Quanzhou, Fujian)

花瓶(浙江嵊州)
Vases (Chengzhou, Zhejiang)

大花瓶(浙江东阳)
A big vase (Dongyang, Zhejiang)

雀目花瓶(福建泉州)
Vases (Quanzhou, Fujian)

双耳旋纹花瓶(福建泉州)　A two-eared vase with spiral grain (Quanzhou, Fujian)

花瓶与奖杯(浙江嵊州)
Vases and prize cup (Chengzhou, Zhejiang)

瓷胎花瓶(四川渠县)
Porcelain-bodied vases (Quxian, Sichuan)

插筋花瓶(浙江新昌)
Vases (Xinchang, Zhejiang)

茶具、酒具、烟具、文具、咖啡具

Tea sets, drinking sets, smoking sets, stationery, coffee sets

瓷胎功夫茶具（四川成都）
A porcelain-bodied tea set (Chengdu, Sichuan)
艺人们高超的编技,在光滑洁白的茶壶、茶杯、托盘上施艺,壶盖、壶柄、杯底及托盘盘沿上,根根篾丝编织到位,给人以清新悦目、精巧细腻的美感。
Bamboo threads are woven accurately on the snow-white fine porcelain teapot, tea cup, tea saucer. All the threads are crossed on the edge of the articles strictly. The whole set is graceful and fascinating

瓷胎茶碟、花瓶（四川成都）
Porcelain-bodied tea saucers and a vase (Chengdu, Sichuan)

瓷胎酒具（四川成都） A porcelain-bodied drinking set (Chengdu, Sichuan)

瓷胎茶具（四川成都）
A porcelain-bodied tea set (Chengdu, Sichuan)

瓷胎功夫茶具(四川成都)
A porcelain-bodied tea set (Chengdu, Sichuan)

瓷胎烟具(四川成都) Porcelain-bodied smoking sets (Chengdu, Sichuan)

瓷胎茶具(四川成都)
A porcelain-bodied tea set (Chengdu, Sichuan)

瓷胎咖啡具(四川成都)
Porcelain-bodied coffee sets (Chengdu, Sichuan)

瓷胎文具(四川渠县)
Porcelain-bodied stationery (Quxian, Sichuan)

瓷胎日式茶具(四川渠县) A porcelain-bodied tea set of Japanese style (Quxian, Sichuan)

瓷胎咖啡具(四川渠县)
A porcelain-bodied coffee set (Quxian, Sichuan)

瓷胎咖啡茶具（四川渠县）
A porcelain-bodied coffee/tea set (Quxian, Sichuan)

瓷胎日式酒具（四川渠县）
Porcelain-bodied drinking sets of Japanese style (Quxian, Sichuan)

瓷胎茶具（四川渠县）
Porcelain-bodied tea sets (Quxian, Sichuan)

瓷胎茶具（四川渠县）
A porcelain-bodied tea set (Quxian, Sichuan)

台屏、字画与龚扇
Table screens, calligraphy, paintings and fans

寿字台屏（四川）
A table screen "Longevity" (Sichuan)

竹编立屏"虎威图"（浙江东阳，卢光华作）
A bamboo-screen stand "Impressive Tiger" (Lu Guanghua, Dongyang, Zhejiang)

瓷胎盘小台屏（四川） Small table screens with a porcelain-bodied plate (Sichuan)

竹编"博古屏风"（浙江东阳，卢光华作）
Bamboo weaving "Ancient Screen" (Lu Guanghua, Dongyang, Zhejiang)

圆盘小台屏与文具（四川）
A small table screen with a round plats, stationery (Sichuan)

竹编双面编小台屏（四川渠县）
Double faced table screens (Quxian, Sichuan)

竹编王羲之书法（四川渠县）
Calligraphy by Wang xizhi (Quxian, Sichuan)

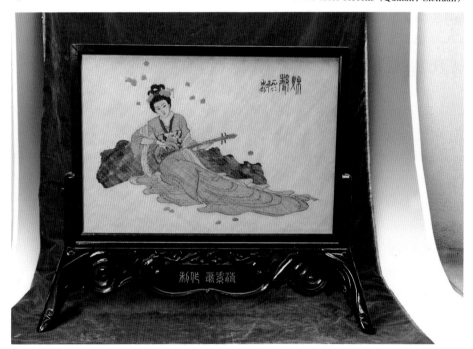
秋声小台屏（四川）
A small table screen "Autumn Sounds" (Sichuan)

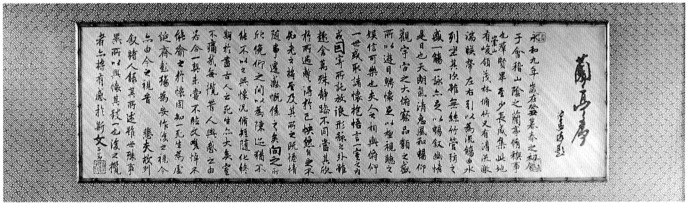
竹编"兰亭序"（浙江东阳，卢光华作）
Calligraphy "Orchid Pavilion" (Dongyang, Zhejiang, Lu Guanghua)

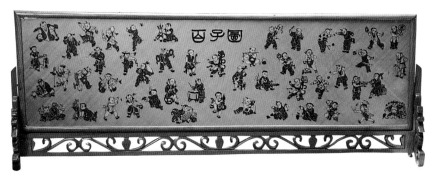

百子图小台屏（四川） A small table screen "One Hundred Sons" (Sichuan)

福与寿台屏（四川） A table screen "Blessedness and Longevity" (Sichuan)

竹编"虎啸图"台屏（四川渠县） A table screen "Roaring Tiger" (Quxian, Sichuan)
创制者以薄如羽翼的篾片，运用提花织物的组织原理，通过同一色泽篾片的经纬交织，将虎的雄风姿态编织在台屏面上。编织面如同锦缎一般，闪闪发亮。由于篾片只有0.002厘米厚，微风可拂，故编织时须用布幔将三方围住，不能有一丝微风才能进行。

Extremely thin strips are used to weave this screen, and silk weaving skills are applied. The vigor and grandeur of tiger are reproduced on the screen, the surface is glittering as brocade. The strips are only 0.002 cm thick, they can be swayed easily in gentle breeze, consequently, the work space must be blocked up with cotton curtain from three sides.

竹编"虎啸图"台屏局部（四川渠县） The detail of the table screen "Roaring Tiger" (Quxian, Sichuan)

竹编"哪吒闹海"台屏（四川渠县） A table screen "A legendary boy Nezha fighting in the sea" (Quxian, Sichuan)

竹编"哪吒闹海"台屏局部(四川渠县) The detail of the table screen "A legendary boy Nezha fighting in the sea" (Quxian, Sichuan)

松鹤旭日台屏(四川)
A table screen "A Pine, a Crane and the Sun" (Sichuan)

竹编字画(云南鲁甸,李祥虎作)
Calligraphy and painting (Ludian, Yunnan, Li Xianghu)

竹编字画(云南鲁甸,李祥虎作)
Calligraphy and painting (Ludian, Yunnan, Li Xianghu)

龚扇(四川自贡)
A Gong fan (Zigong, Sichuan)

奔马小台屏(四川渠县) A small table screen "Running Horse" (Quxian, Sichuan)

竹编宫灯（福建泉州） A bamboo woven lamp of palace style (Quanzhou, Fujian)

龚扇（四川自贡）
A Gong fan (Zigong, Sichuan)

竹编宫灯（福建泉州） A bamboo woven lamp of palace style (Quanzhou, Fujian)

龚扇（四川自贡） A Gong fan (Zigong, Sichuan)
四川自贡的龚扇，是闻名中外的竹丝扇，造型古朴大方，选料精细讲究，扇面薄如纨绢，几近透明，其编织之精细堪称中国之最。这种扇由龚氏家族世代相传制作，故称"龚扇"。
Gong fan is produced in Zigong City of Sichuan Province, which is a kind of well-known bamboo filament fan. The shape is simple and in good taste. The selection of raw material is fastidious. The fan is as thin as silk and nearly transparent. Its fine workmanship is the best in China. This kind of fan is produced by Gong's family from generation to generation, so it is called "Gong Fan".

龚扇（四川自贡）
A Gong fan (Zigong, Sichuan)

灯具 Lamps

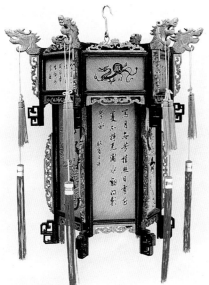

竹编宫灯(四川渠县) A bamboo woven lamp of palace style (Quxian, Sichuan)

竹编灯具(浙江新昌) Bamboo woven lamps (Xinchang, Zhejiang)

蟠龙竹丝台灯(福建泉州) A bamboo woven lamp with a coiling dragon (Quanzhou, Fujian)

竹编篮与箱(浙江新昌) Bamboo woven baskets and boxes (Xinchang, Zhejiang)

竹编罐(浙江东阳) Bamboo woven pots (Dongyang, Zhejiang)

蝴蝶花插(福建泉州) A butterfly flower holder (Quanzhou, Fujian)

罐、筐、篓、包及其它 Pots, baskets, boxes, etc.

竹编漆器罐（福建泉州） Lacquered bamboo woven pots (Quanzhou, Fujian)

竹编筐与篓（福建泉州）
Bamboo woven baskets (Quanzhou, Fujian)

竹编渔篓（广东陆丰） Bamboo woven fishing baskets (Lufeng, Guangdong)

竹编罐与瓶（浙江新昌）
Bamboo woven pots and a bottles (Xinchang, Zhejiang)

蝉壁插（浙江嵊州）
A cicada flower holder (Chengzhou, Zhejiang)

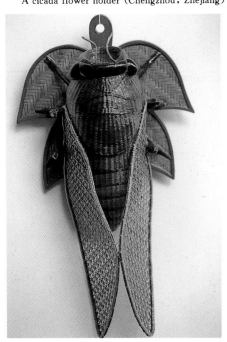

烫金插筋花瓶与鹅篮（浙江嵊州） Vase and a goose-shaped baskets (Chengzhou, Zhejiang)

仿禽鸟类的竹篮深受消费者喜爱，其中尤以 鹅篮 为最，鹅身的"编"，鹅翅的"插"，鹅掌的"弹"，巧妙地结合，而且大小数只，可以迭套，便于运输。花瓶高21厘米，直径最大处8.7厘米。

The bird-imitated bamboo baskets, especially the Goose Basket, is keenly loved by consumers. Different weaving skills have been ingeniously integrated, and the products of different sizes are designed to be put one into another, which can be packed and transported very conveniently. This vase is 21 cm in height, its maximum diameter is 8.7 cm.

扇形篓（福建泉州）
Fan-shaped baskets (Quanzhou, Fujian)

精制竹编斗笠（福建顺昌） Highly finished bamboo hat (Shunchang, Fujian)

拖鞋（四川成都）
Slippers (Chengdu, Sichuan)

竹夫人（江西瑞金） A bamboo article for cooling on bed (Ruijin, Jiangxi)

鼎（浙江东阳）
An incense burner with three legs (Dongyang, Zhejiang)

笠帽（浙江东阳）
A bamboo cap (Dongyang, Zhejiang)

南方竹制用品
Bamboo handicrafts from the South

提包（四川）
Hand-bags (Sichuan)

公鸡（浙江嵊州）Cock (Shengzhou, Zhejiang)
由俞樟根创制的第一只竹编禽鸟（1967年）。
The first bamboo woven fowl made by Yu Zhanggen.

鸳鸯（浙江安吉，徐华铛、陈庆良创作）
Mandarin ducks (created by Xu Huadang, Chen Qingliang, Anji, Zhejiang)
中国竹子博物馆（安吉）赠送日本朋友的礼品。
A gift made by Chinese Bamboo Museum (Anji) to Japanese friends.

鸬鹚（浙江嵊州）
A cormorant (Shengzhou, Zhejiang)

喜鹊登梅（浙江嵊州）
Magpies on a plum (Shengzhou, Zhejiang)

白孔雀(浙江嵊州)
A white peacock (Shengzhou, Zhejiang)

双头锦鸟盘（浙江东阳） A plate with a golden pheasant (Dongyang, Zhejiang)

小飞鹰（浙江嵊州）
A flying eagle (Chengzhou, Zhejiang)

顶饰鹤（浙江嵊州）
A crane (Chengzhou, Zhejiang)

山鹰展翅（浙江嵊州）
A eagle spreading the wings (Chengzhou, Zhejiang)

白尾海雕（浙江嵊州） A white-tailed eagle (Chengzhou, Zhejiang)
该作品用篾片仿照海雕的长短羽毛编插，生动自然。海雕高58厘米，神形酷肖，双爪得体地栖息在天然质朴的毛竹箬头上，体现了竹的属性。该海雕于1979年初进入美国白宫。
Bamboo strips were used to simulate the plumage of eagle incisively and vividly. The eagle is 58 cm in height. It is alike both in appearance and spirit. The two talons of the eagle are put on the top of bamboo which also show the character of bamboo. This handicraft was collected by White House in 1979．

群鸡(浙江嵊州)
Cocks (Chengzhou, Zhejiang)

鸳鸯(浙江嵊州)
Mandarin ducks (Chengzhou, Zhejiang)

猫头鹰展翅(浙江新昌) An owl spreading the wings (Xinchang, Zhejiang)

猫头鹰(浙江嵊州)
An owl (Chengzhou, Zhejiang)

锦鸡(浙江嵊州)
A golden pheasant (Chengzhou, Zhejiang)

天鹅(浙江嵊州)
Swans (Chengzhou, Zhejiang)

鹦鹉(浙江东阳)
A parrot (Dongyang, Zhejiang)

双鹅(浙江东阳)
Goose (Dongyang, Zhejiang)

鸡盘与鹅盘(浙江嵊州) A cock-shaped plate and goose-shaped plates (Chengzhou, Zhejiang)

白鸡盘与白鹅盘(浙江嵊州) White cock-shaped plates and white goose-shaped plates (Chengzhou, Zhejiang)

鸭盘与雁盘(浙江嵊州) Duck-shaped plates and wild-goose-shaped plates (Chengzhou, Zhejiang)

鹅盘(浙江嵊州)
Goose-shaped plates (Chengzhou, Zhejiang)

金鸡报晓(浙江东阳)
A golden cock as a harbinger of dawn (Dongyang, Zhejiang)

双头鹅篮(浙江东阳) Double-headed goose-shaped baskets (Dongyang, Zhejiang)

雄狮（浙江东阳，何福礼）
Lion (He Fuli, Dongyang, Zhejiang)

狼犬（浙江横店，中国竹编博物馆）
Wolfhound (Hengdian, Zhejiang, Chinese Bamboo Weaving Museum)

关爱（浙江东阳）
Solicitude (Dongyang, Zhejiang)

麻鸭（浙江嵊州，张樟祥作）
Duck (Zhang Zhangxiang, Shengzhou, Zhejiang)

扬鼻象（浙江东阳，何福礼作）
An elephant raising proboscis (He Fuli, Dongyang, Zhejiang)

盘羊（浙江嵊州）
Sheep (Shengzhou, Zhejiang)

仙鹤（浙江嵊州，胡六久作）
Legendary crane (Hu Liujiu, Shengzhou, Zhejiang)

狮面鱼（浙江嵊州）
A fish with a lion face (Shengzhou, Zhejiang)

双飞鹤（浙江嵊州）
Twin flying cranes (Shengzhou, Zhejiang)

行马（浙江嵊州）A running horse (Shengzhou, Zhejiang)
浙江的竹编有中国的民族特色。竹编"行马"高37厘米，长38厘米。
Zhejiang bamboo weaving demonstrates distinctive Chinese national features. This article is 37cm in height and 38 cm in length.

灰鹅盘（浙江嵊州）
Grey goose-shaped plates (Chengzhou, Zhejiang)

鸡盘与鹅盘（浙江嵊州） A cock-shaped plate and a goose-shaped plate (Chengzhou, Zhejiang)

圆盒（四川）
Round boxes (Sichuan)

公鸡篮与猫头鹰篮（浙江嵊州） Cock-shaped basket and owl-shaped baskets (Chengzhou, Zhejiang)

鹅篮与鸭篮（浙江嵊州）
Goose-shaped baskets and Duck-shaped baskets (Chengzhou, Zhejiang)

凤凰（浙江新昌）
Phoenix (Xinchang, Zhejiang)

火鸡篮（浙江东阳）
Turkey-shaped baskets (Dongyang, Zhejiang)

镂空鹅篮（福建泉州） Pierced goose-shaped plates (Quanzhou, Fujian)

禽鸟器皿（浙江新昌）
Bird-shaped service (Xinchang, Zhejiang)

鸡篮与鸡罐（广东南雄） A cock-shaped basket and cock-shaped pot (Nanxiong, Guangdong)

鸡盘与鹅盘（浙江嵊州） Cock-shaped plates and goose-shaped plates (Chengzhou, Zhejiang)

待运的公鸡罐（浙江嵊州）
Cock pots to be shipped (Chengzhou, Zhejiang)

禽篮一组（浙江新昌） A set of bird-shaped baskets (Xinchang, Zhejiang)

狗与猫头鹰（浙江东阳）
A dog and a owl (Dongyang, Zhejiang)

仙鹤瓶（浙江嵊州）
A crane pot (Chengzhou, Zhejiang)

小动物罐（福建泉州）
Animal pots (Quanzhou, Fujian)

飞鹰（浙江新昌）
Flying eagles (Xinchang, Zhejiang)

小动物与插花篓（广东陆丰）
Small animals and a flower holder (Lufeng, Guangdong)

走兽 Beasts

大象与小象(浙江嵊州)
Elephants (Chengzhou, Zhejiang)

母子象(浙江东阳)
Elephants (Dongyang, Zhejiang)

大象（浙江东阳）
Great elephant (Dongyang, Zhejiang)
作品长430厘米，高216厘米，编织严密完整，造型自然生动，是目前我国竹编动物中最大的作品。获首届国际（杭州）民间手工艺品博览会金奖。
The works is 430 cm in length and 216 cm in height, the woven structure is compact and the contour is vivid. This is the biggest bamboo woven article in China at present, winner of a gold prize at international fair (Hangzhou) of folk handicrafts.

大熊猫（浙江东阳）
Giant pandas (Dongyang, Zhejiang)

群象（浙江嵊州）
Elephants (Chengzhou, Zhejiang)

犀牛（浙江东阳）
Rhinoceros (Dongyang, Zhejiang)

奔马（浙江东阳）
A running horse (Dongyang, Zhejiang)

牛斗虎（浙江东阳）
An ox fighting a tiger (Dongyang, Zhejiang)

群虎（浙江嵊州）
Tigers (Chengzhou, Zhejiang)

蛇（福建泉州）
A snake (Quanzhou, Fujian)

群马（浙江嵊州）
Horses (Chengzhou, Zhejiang)

母子虎（浙江东阳）
Tigers (Dongyang, Zhejiang)

斗牛局部（浙江新昌）
The detail of bull-fighting (Xinchang, Zhejiang)

斗牛（浙江新昌）
Bull-fighting (Xinchang, Zhejiang)

大象（浙江乐清，王应松作）
Elephants (Leqing, Zhejiang, Wang Yingsong)

西班牙斗牛（浙江东阳）
Spanish bull-fighting (Dongyang, Zhejiang)

骆驼与斑马（浙江东阳）
Camels and a zebra (Dongyang, Zhejiang)

黄牛与奶牛（浙江东阳）
An ox and a cow (Dongyang, Zhejiang)

牛与小车(浙江东阳)
An ox and a cart (Dongyang, Zhejiang)

三羊开泰(浙江嵊州)
Three goats bring happiness (Chengzhou, Zhejiang)

奶牛与黄牛(浙江嵊州)
Cows and an ox (Chengzhou, Zhejiang)

动物一组(浙江新昌)
Animals (Xinchang, Zhejiang)

哈巴狗(浙江嵊州)
Pekingeses (Chengzhou, Zhejiang)

哈巴狗(浙江嵊州)
Pekingeses (Chengzhou, Zhejiang)

母子鹿(浙江嵊州)
Deers (Chengzhou, Zhejiang)

母子牛(浙江嵊州)
A cow with carves (Chengzhou, Zhejiang)

奶牛（浙江东阳）
A cow (Dongyang, Zhejiang)

金鱼壁挂（台湾台北）
Wall hanging "Gold Fish" (Taipei, Taiwan)

马踏飞燕（浙江东阳）
Horses step on a swallow (Dongyang, Zhejiang)

熊猫罐（浙江嵊州）
A pot "Panda" (Chengzhou, Zhejiang)

牦牛罐（江西）
A pot "Yak" (Jiangxi)

猪（浙江东阳）
Pigs (Dongyang, Zhejiang)

小门狮与三脚鼎（浙江嵊州）
Small lions and a incense burner with three legs (Chengzhou, Zhejiang)

哈巴狗（浙江东阳）
Pekingeses (Dongyang, Zhejiang)

羊（浙江东阳）
Goats (Dongyang, Zhejiang)

翠竹飞鸟（浙江嵊州）　Green bamboos and flying birds (Chengzhou, Zhejiang)

大门狮（浙江嵊州）
Great lions (Chengzhou, Zhejiang)

灵禽神兽　Legendary animals and birds

沧海还珠（浙江省嵊州市工艺竹编厂集体创作）
Playing with a pearl from the sea (Bamboo Weaving Factory, Shengzhou City, Zhejiang Province)
浙江省政府在澳门回归祖国之际赠送澳门特区政府的礼品。
A gift made by Zhejiang Provincial Government to Macao Special Administrative Region rejoining to motherland.

千禧龙(浙江横店,中国竹编博物馆,何福礼等作)
Dragon of New Century (made by He Fuli et al, Hengdian, Zhejiang, Chinese Bamboo Weaving Museum)

龟（浙江嵊州）Tortoises (Shengzhou, Zhejiang)

群鹿（浙江东阳）Deers (Dongyang, Zhejiang)

小兔拉车（福建泉州）Rabbits and a cart (Quanzhou, Fujian)

玉麒麟（浙江嵊州）
Jade lion (Shengzhou, Zhejiang)

哪吒闹海（浙江东阳，何福礼、何红兵、何红亮、蔡平义作）
Nezha, a divine warrior, conguering a dragon on the Sea (made by He Fuli, He Hongbing, He Hongliang, Cai Pingyi)

双龙戏珠壁挂（福建泉州）
A wall hanging "Two dragon playing a pearl" (Quanzhou, Fujian)

狮子滚绣球（浙江东阳）
Lions playing a ball (Dongyang, Zhejiang)

蟠龙壁挂（福建泉州） A wall hanging "A Coiling Dragon" (Quanzhou, Fujian)

蟠龙壁挂（福建泉州） A wall hanging "A Coiling Dragon" (Quanzhou, Fujian)

麻姑献寿（浙江东阳）
The fairy maiden Magu congratulates on a birthday ceremony (Dongyang, Zhejiang)

龙凤盘（浙江嵊州）
A plate "A Dragon and a Phoenix" (Chengzhou, Zhejiang)

圣诞老人（浙江嵊州）
Santa Claus (Chengzhou, Zhejiang)

龙凤灯架（浙江东阳） A lamp holder "A Dragon and a Phoenix" (Dongyang, Zhejiang)

小龙舟（浙江嵊州） A small dragon boat (Chengzhou, Zhejiang)

九狮舞绣球（浙江嵊州） Nine Lions and a silk balll (Chengzhou, Zhejiang)
作品长 90 厘米，高 70 厘米，宽 60 厘米。现陈列于北京人民大会堂浙江厅。
It is 90 cm in length, 70 cm in height and 60 cm in width. Now it is on display in Zhejiang Room of the Great Hall of the People.

镂空九龙之一（福建泉州）
Pierced dragons 1 (Quanzhou, Fujian)

镂空九龙之二（福建泉州）
Pierced dragons 2 (Quanzhou, Fujian)

镂空九龙之三（福建泉州）
Pierced dragons 3 (Quanzhou, Fujian)

镂空九龙之四（福建泉州）
Pierced dragons 4 (Quanzhou, Fujian)

镂空九龙之五（福建泉州）
Pierced dragons 5 (Quanzhou, Fujian)

镂空九龙之六（福建泉州）
Pierced dragons 6 (Quanzhou, Fujian)

镂空九龙之七(福建泉州)
Pierced dragons 7 (Quanzhou, Fujian)

龙凤呈祥台屏与斛(浙江东阳)
A table screen "A Dragon and a Phoenix bringing Prosperity" (Dongyang, Zhejiang)

镂空九龙之八(福建泉州)
Pierced dragons 8 (Quanzhou, Fujian)

镂空九龙之九(福建泉州)
Pierced dragons 9 (Quanzhou, Fujian)

龙凤壁挂(福建泉州)
A wall hanging "A Dragon and a Phoenix" (Quanzhou, Fujian)

一般动物编织都有一个胎模支撑,艺人们只要依胎编织即可。而该作品不仅造型生动,而且九条龙运用了串丝镂空的特技,确是竹编工艺中的一绝。
Ordinary animal-shaped articles are woven each on one model body. But the weaving skill of this article is much more complicated. The pierce weaving of stringed threads is the most difficult one in bamboo weaving.

历史神话人物 Legendary and historical stories

麻姑献寿（浙江嵊州） The fairy maiden Magu congratulates on a birthday ceremony (Chengzhou, Zhejiang)

苏武牧羊（浙江嵊州） The envoy shepher Su Wu (Chengzhou, Zhejiang)

渔翁（浙江东阳） A fisherman (Dongyang, Zhejiang)
该作品高310厘米，是我国目前竹编人物中最大的。艺人们巧妙地变换经丝和纬丝，把整个人物形象编得严密完整，浑然一体。在文化部举办的全国民间艺术大赛中获金奖。
It is 310 cm in height, which is the largest bamboo-woven figure up to date. The artists change ingeniously the warp and weft bamboo strips and weave the figure into a unified entity. It won the gold prize of folk arts competition organized by the Ministry of Culture.

武松打虎（浙江新昌）
The hero Wu Song fighting a tiger (Xinchang, Zhejiang)

大龙（福建泉州） A great dragon (Quanzhou, Fujian)

建筑艺品 Architecture

麒麟送子（浙江东阳） The Kylin brings along a son (Dongyang, Zhejiang)

弥勒佛（浙江东阳） The master Bo Le studies horses (Dongyang, Zhejiang)

岳飞（浙江嵊州） Yue Fei (Chengzhou, Zhejiang)

作品溶各种编织技法于一炉，除竹丝编织外，还应用了笋壳贴、翻簧等工艺。长110厘米，高116厘米，曾多次出国展出，在1980年举行的"中国草柳藤竹编织交易会"上，这件精品被一致公认为是当时最高档竹编品。

This work of art combined various weaving techniques. Bamboo shell pasting and "Fanhuang" techniques were used beside bamboo strip weaving. It is 110 cm in length and 116 cm in height. It has been exhibited abroad for many times. In "China Fair of Willow and Bamboo Weaving" held in 1980, this fine work of art was generally acknowledged as the top-class bamboo-woven handicraft up to date.

花塔（浙江嵊州） The Flower Pagoda (Chengzhou, Zhejiang)

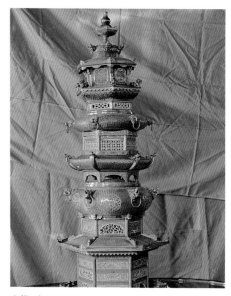

宝塔（浙江东阳） The Great Pagoda (Dongyang, Zhejiang)

"秦始皇"壁挂（浙江东阳） A wall hanging "Emperor Qin Shihuang" (Dongyang, Zhejiang)

老寿星（浙江新昌） The God of Longevity (Xinchang, Zhejiang)
该作品是人物编织中的精品，着力刻划了寿星硕大的前额和慈祥的神态。作品高96厘米，宽57厘米。在1986年全国工艺美术品百花奖评比会上获"希望杯"称号。
This is a fine work of art of bamboo woven figure. It concentrates the effort on portraying the god's big forehead and kindly manner. It is 96 cm in height, 57 cm in width. This work of art won the "Hope Cup" in national arts and handicrafts competition in 1986.

老寿星局部（浙江新昌）
The detail of the God of Longevity (Xinchang, Zhejiang)

秦陵铜车马
该作品原是秦始皇陵的陪葬品，现通过竹编再现了原物的风采，作品长140厘米，高62厘米，1989年秋，在中国工艺美术馆落成时，参加了全国工艺美术品展览，专家们对运用各种编织法去体现铜车原来图纹的装饰，深表赞叹。
"Bronze Horses and Chariot" of Qin Tomb
Bronze Horses and Chariot was unearthed from the Qin Tomb. This bamboo woven article reproduces the elegance of the bronze original. It was displayed on the National Exhibition of Handicrafts on the occasion of inauguration of Chinese Handicrafts and Arts Hall in 1989, and was highly praised by the professional experts.

秦陵铜车马局部之一（浙江嵊州）
The detail of the bronze horse and cart unearthed from Qin Tomb 1 (Chengzhou, Zhejiang)

秦陵铜车马（浙江嵊州）
The bronze horse and cart unearthed from Qin Tomb (Chengzhou, Zhejiang)

秦陵铜车马局部之二（浙江嵊州）
The detail of the bronze horse and cart unearthed from Qin Tomb 2 (Chengzhou, Zhejiang)

工艺竹亭（福建顺昌）　Bamboo tavilion
(Shunchang, Fujian)

大龙舟(浙江嵊州)
The great dragon boat (Chengzhou, Zhejiang)
该作品把龙身和亭台楼阁巧妙地结合在一起,造型别致,装饰豪华。全长 196 厘米,高 152 厘米,是中国竹编龙舟之最。在 1989 年第六届全国工艺美术品百花奖上被评为创新设计一等奖。
The body of dragon and pavilion were ingeniously combined together. The shape is special and the decorate is luxurious. It is 196 cm in length and 152 cm in height, which is the largest bamboo-woven dragon boat in China. It won the first prize of creative designing in the national arts and handicrafts competition in 1989.

大龙舟局部(浙江嵊州)
The detail of the great dragon boat (Chengzhou, Zhejiang)

孙悟空三打白骨精（浙江东阳） A legendary monkey fighting a ghost (Dongyang, Zhejiang)

伯乐相马（浙江东阳） The master Bo Le stadies horses (Dongyang, Zhejiang)

天坛（浙江新昌） Temple of Heaven (Xinchang, Zhejiang)

该作品以富丽壮观、场面宏大取胜。高203厘米，直径210厘米，艺人们运用多种编织技法，并配以贴花、扎藤、竹雕、翻簧、镶嵌等综合工艺，充分体现了天坛庄严、典雅、宏伟的古建筑风貌。

This masterpiece is celebrated for its magnificence, it is 203 cm in height, and the diameter is 210 cm. Many weaving techniques were used. The solemnity, elegance and magnificence of this ancient building is fully expressed.

越乡古戏台(浙江嵊州)
An ancient stage in a Zhejiang village (Chengzhou, Zhejiang)

香炉阁(浙江东阳) The Incense Burner Pavilion (Dongyang, Zhejiang) 该作品的底座是三脚大香炉鼎,上面高耸着玲珑精致的双层八角楼阁,顶端装饰着洁白昂首的仙鹤,显得挺拔高雅。作品高150厘米,现陈列在北京人民大会堂浙江厅。

The base of this handicraft is a tripod incense burner, on which a two-story pavilion stands. The top is decorated with white cranes. It looks very elegant and noble. It is 15 cm in height. Now it is on display in Zhejiang Room of the Great Hall of the People.

六和塔(浙江嵊州) The pagoda "Liuhe Tower" (Chengzhou, Zhejiang) 作品参照原塔的造型进行竹编的再创造,分塔身和底座两部分,高66厘米,直径35厘米,揭开塔尖顶盖,里面竟是一个内壁光洁的容器,从而使这件精美的工艺品增添了实用价值。

The artists consulted the original shape of Liuhe Tower and wove it by bamboo strips. This work of art is composed of tower body and base. It is 66 cm in height and 35 cm in diameter. Uncovering the top of tower, you will find that it is a container with smooth internal wall. So it is not only beautiful but also practical.

牡丹亭(浙江嵊州) The peon pavilion (Chengzhou, Zhejiang)

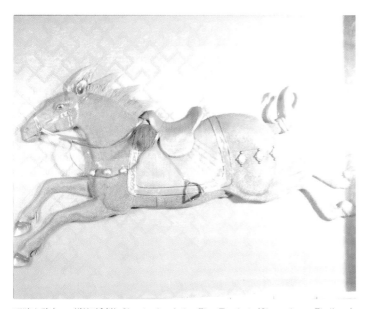
昭陵六骏之一（浙江嵊州）Six steeds of the ZhaoTomb 1 (Shengzhou, Zhejiang)

昭陵六骏之二（浙江嵊州）Six steeds of the ZhaoTomb 2 (Shengzhou, Zhejiang)

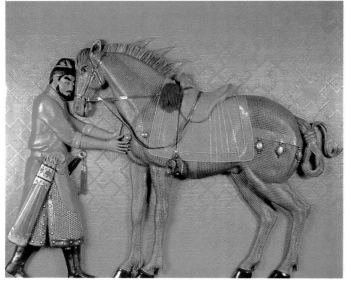
昭陵六骏之三（浙江嵊州）Six steeds of the ZhaoTomb 3 (Shengzhou, Zhejiang)

昭陵六骏之四（浙江嵊州）Six steeds of the ZhaoTomb 4 (Shengzhou, Zhejiang)

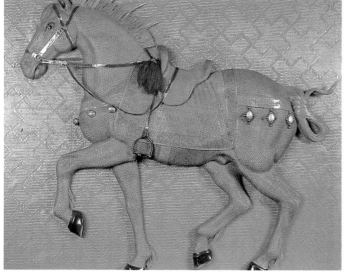
昭陵六骏之五（浙江嵊州）Six steeds of the ZhaoTomb 5 (Shengzhou, Zhejiang)

昭陵六骏之六（浙江嵊州）Six steeds of the ZhaoTomb 6 (Shengzhou, Zhejiang)

九龙壁（浙江东阳）　Nine dragon wall (Dongyang, Zhejiang)

编织精工绝伦,色泽素净高雅,显示竹篾本色。长6.19米,高2.68米,底座宽0.55米。在1984年举行的第四届中国工艺美术品百花奖评审会上,被专家们评为百花奖珍品,同时授予金杯奖,是目前全国竹编行业中获得的最高殊荣。

The weaving technique is exquisite. The color and lustre is simple but elegant and shows the natural characters of bamboo strips themselves. It is 619 cm in length and 268 cm in height. The width of base is 55 cm. "Nine Dragons Wall" was awarded a gold prize cup as well as the title of "art treasure" of Hundred Flowers Prize in 1984 which is the highest honor ever won by bamboo weaving profession.

九龙壁局部之一（浙江东阳）　The detail of nine dragon wall, 1 (Dongyang, Zhejiang)

九龙壁局部之二（浙江东阳）　The detail of nine dragon wall, 2 (Dongyang, Zhejiang)

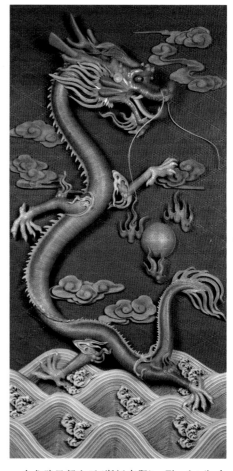

九龙壁局部之三（浙江东阳）　The detail of nine dragon wall, 3 (Dongyang, Zhejiang)

九龙壁局部之四（浙江东阳） The detail of nine dragon wall, 4 (Dongyang, Zhejiang)

九龙壁局部之五（浙江东阳） The detail of nine dragon wall, 5 (Dongyang, Zhejiang)

九龙壁局部之六（浙江东阳） The detail of nine dragon wall, 6 (Dongyang, Zhejiang)

九龙壁局部之七（浙江东阳） The detail of nine dragon wall, 7 (Dongyang, Zhejiang)

九龙壁局部之八（浙江东阳） The detail of nine dragon wall, 8 (Dongyang, Zhejiang)

九龙壁局部之九（浙江东阳） The detail of nine dragon wall, 9 (Dongyang, Zhejiang)

THE VARIETY OF ART AND CRAFTS IN BAMBOO REALM

竹笵雜小藝

以刀代笔，在竹筒、竹片、竹根上进行施艺，可雕刻成山水、亭、台、楼、阁、人物、花草、树木、飞禽等各种艺术造型。文人用的笔筒、笔床、臂搁以及扇骨、香筒、联对、竹杖、插屏、佛手、鸟笼，甚至妇女头上的簪钗等都有竹刻的饰物。

用刀刻竹，在中国古时早已发明。中国在发明造纸术以前，以"竹简"作为文字的载体。古籍《论衡·量知》称："截竹为筒，破以为牒，加笔墨之迹，乃成文字"。湖南长沙马王堆一号西汉墓出土的彩漆竹勺，勺柄以龙纹及辫纹为饰，是迄今发现中国最早的竹刻艺术品。唐代竹刻渐趋艺术化，至宋代更加进步。据记载，宋代詹成能在竹片上雕刻宫室、山水、人物、花鸟，纤毫具备。明、清时期，竹刻成为一种专门艺术，竹刻名家迭起，技术越发成熟，争相传习，成为历史上的辉煌时期。

竹刻的用材，大多选用径粗质坚、节间长、通直、无病虫害的毛竹，以3～4年生老竹为宜。雕刻笔筒、扇骨之类应截取竹杆；进行人物、鸟兽等的造型则应选取竹根。选用的竹材、竹根必须经熏蒸、干透才可施艺。

竹刻工具简单，只要有锯子和刻刀就可操作。一般刻刀用平口刀，也兼用圆头刀、凹卷刀，刻刀无定型式样，根据需要自行磨制，以便于操作为宜。

竹刻的刀法，根据奏刀的深浅、方式不同分为阴刻和阳刻。阴刻又有线刻、浅刻、深刻等；阳刻有留青、薄地阳文、浅浮雕、高浮雕、透雕等。线刻刻痕细如发丝，常见于画面的细部刻划，如人物的眉毛、胡须等；浅刻刻纹较浅，有线有面，表现丰富，如刻植物的花瓣、绿叶等；深刻，即所刻较深，能洞穿竹之肌肉，应用甚广，为最基本之刻法。留青，称皮雕，即留取竹表面青色部分作施艺部位，而将竹之肌肤作为底色，是竹刻的一种重要技法。浮雕，即在竹之表面刻成浅立体形象，无论山水、人物、动物都可用浮雕表现出来，浮雕也有深、浅之分，深的称为高浮雕，浅的称为浅浮雕，薄地阳文实际上是一种浅浮雕，所刻花纹微微隆起；透雕，即把竹肌刻穿，进行镂空造型；圆雕，则刻出完整的立体形状，多以竹根作施艺材料。

竹刻作品的题材比较广泛，大多取材山水、花草、树木、飞禽以及人物、

老寿星　（南京，羊光作）　God of Longevity (Nanjing, Yang Guang)

仙佛、道释等。书画为竹刻之基础，竹刻艺人往往能书善画，或精于篆刻印章，巧于艺术构思，运用娴熟的竹刻技法而成佳作。有的所刻山川云树迂曲盘折，生动自然；有的所刻花草玲珑有致，神形毕肖；有的所刻山水楼阁工细绝伦，气象万千；有的所刻人物，细腻传神，衣纹缥缈；有的又完全根据竹之盘根错节，略加刮磨即已成器，这全在于艺人的艺术造诣和娴熟的刀法。

本画册选载的竹刻作品，也可窥见中国竹刻艺术之风采。明代竹刻大师朱松邻所刻的《松鹤笔筒》，作品布局完整、生动自然，雕一段老松巨干，密布鳞片瘿节，另出一松枝，迂曲环抱，针叶繁茂，松畔双鹤各具姿态。系为人祝寿之作。朱松邻，名鹤，为嘉定派竹刻创始人，以深刻作高浮雕著称，精于笔筒，又长于簪钗，所刻笔筒、香筒、杯、簪钗等诸器为世人之宝，传世作品极为罕见，此件为南京博物院收藏。明代竹刻名家濮仲谦落款的"八仙

竹刻 Bamboo engraving

"孙子兵法"竹简　（南京，羊光作）　Bamboo tube "The Art of War" by Sunzi (Nanjing, Yang Guang)
该作品由700多块竹简组成，作者运用了留青刻法。留青是竹刻艺术中最难的一种，它是利用竹子表面的竹青雕刻出艺术形象，将空白处竹青铲去露出的肌层作为底色，可见其工之精微。
This is made of some seven hunrdred pieces of bamboo meterial, applying green engraving techniques. The green surface of bamboo material is used to present the artificial image, while the bamboo grain in vacant areas are used as the background, the operation is quite difficult.

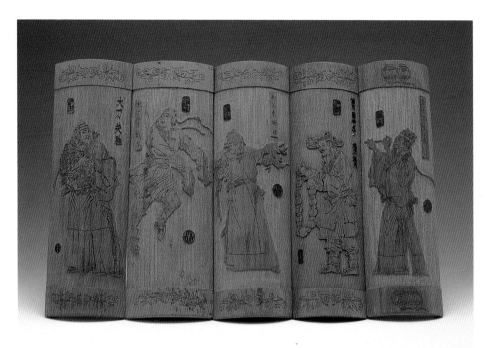

水浒人物 （南京，羊光作）
Heros of the Marshes (Nanjing, Yang Guang)

Landscape, architecture, figures, flowers, grasses, trees and birds can be carved on bamboo slips, tubes and bamboo roots with a buirn instead of a pen (brush), resulting in variety of artistic shaping. The pen containers, antithetical couplets, walking sticks, table plasques, Buddha's hand, bird cages, and even hairpins all show traces of bamboo sculpture.

Bamboo sculpture with a burin has long been invented in ancient China. Before the invention of paper making, bamboo slips were used as carriers of Chinese characters. In the ancient classics: "A thesis on criteria of judgment: assessment of knowledge" (Wang Chong's "Lunheng") it was stated: Culm is cut into tubes on which characters were written with ink." The bamboo ladles and ladle handles coated with colored lacquer, unearthed from the No.1 West Han tomb at Mawangdui in Changsha show decorative patterns of dragons and braids. They are the earliest art objects of bamboo sculpture ever discovered in China. In Tang Dynasty the bamboo sculpture tended to have an artistic quality. In Song Dynasty much more progress was made. Historical records show that in Song Dynasty the palaces,

笔筒"，刻悬崖峭壁，浓云密布，仙人乘木筏于急湍波涛之上而泰然自的情景。整个画面深浅高低、层次分明、错落有致、气势非凡，是竹刻珍品。明代竹刻高手芷岩竹刻《山水纹笔筒》，呈暗红色，用留青技法，所刻山水风景，线条流畅，层次分明，很有特色。

台湾台北《康庐竹缘》的竹刻家李佑增所刻之联对，刀法精湛，笔力遒劲，在再现了书法原貌的基础上有所创新，成为称誉台湾岛的竹刻大家。湖南衡东谭南森刻《五百罗汉》，长1400厘米，高54厘米，艺人以娴熟的刀工，在216块竹简上镂刻了五百罗汉形象，组成一幅形象明显，呼应连贯，气势恢弘的艺术长卷，系目前中国大型竹刻品。浙江桐乡竹刻《荷花》《书法》系中国目前竹刻中的精品。浙江象山的竹根雕在我国是很有名气的，这里名家荟萃，技艺超群，有"中国竹根雕之乡"之美称。这里我们撷选了几件精品，如竹根雕《张飞》，利用须根恰到好处地刻成了张飞的眉毛、须发和胡子，脸部突出强化张飞的神态，豹头环眼、燕领虎须，一脸怒气，似能听到他巨雷般的吼声。另一件根雕人物头像《沉思》，人物造型深沉刚毅，面部肌肉凸起处似有弹性。竹根雕《小伙伴》形肖神似。广东梅县竹刻《足球之乡》笔筒，全部镂空刻成，独具匠心。南京竹刻《孙子兵法》竹简，运用阳刻技法，在300张竹片上镌刻近6000字，字字俊秀，是竹刻之大件。

用竹枝、竹筒、竹片、竹节造型的竹玩具在中国南方旅游景点到处可见，很受人们的喜爱。艺人们利用竹枝、竹节等自然形态，根据自己的艺术构思，经锯切、镶嵌等手法创造形象，在似与非似之间给人以丰富的想象。如广东南雄的《竹帆船》《飞跃》《双鹿》等充满乡土气息的作品，很具表现力。最使人新奇的是取嫩竹尖上的细小的竹管为原料制成的竹衣，其表面呈网格状，穿着凉爽又富新意。

用竹笋壳贴制的人物形象更为传神。浙江乐清制作的历史人物《关公》，相貌堂堂、威风凛凛，把书上描写的美髯公形象表现的维妙维肖。《济公》《寿星》等也别具神韵。

除以上谈及的竹刻、竹雕、竹枝、竹筒、竹片造型、笋壳贴制人物外，还有竹杖、竹筷、竹乐器、竹笔杆、竹伞、竹扇、风筝、灯彩等竹制用品，大家可沿着我们所摄制的照片，去领略一番竹苑杂艺的风采。

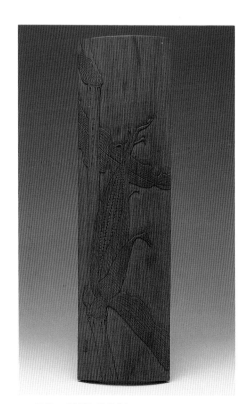

玉米 （南京，羊光作）
Maize (Nanjing, Yang Guang)

透雕笔筒 （南京，羊光作） A thoroughly carved brush pot (Nanjing, Yang Guang)

landscape, figures, flowers and birds carved on bamboo slips by Zhen Changneng were close to reality in details. In Ming and Qing Dynasties bamboo sculpture has become a special art, and there appeared a galaxy of famous bamboo sculptors. The art has become sophisticated. Many emulated in learning it and extending it. Ming and Qing Dynasties were the magnificent period of bamboo sculpture.

Most of the materials used is the straight, long-internode, thick moso bamboo which should be 3-4 years old and be free from diseases and insects. Culms are used for making pen container, and fan ribs. Bamboo roots are used in artistic shaping of figures, birds and animals. The bamboo culms and roots must be fumigated and thoroughly dried before being sculptured.

The operating tools are simple. Besides a saw, there are burins. The general burin has a flat blade, but it may have a round head or a concave blade. No fast rule is laid down for its shape. Its shape depends on special use. In short the shape should suit the need of flexible operation.

The method of sculpture is divided into concave engraving and convex carving. In accordance with the depth and mode of cutting, the concave engraving is subdivided into linear, shallow and deep engravings, while the convex carving into green carving, shallow convex carving, shallow relief, high relief and through carving. The line of linear engraving is as fine as a hair, often seen in details of a work, such as a persons' eyebrow and beard. The line of shallow engraving is relatively shallow — lines being connected with areas, noted for expressiveness such as floral petals and green leaves of plants. Deep engraving implies deep cut or piercing of the bamboo flesh. Due to wide use, it is the most important fundamental method. Green carving is also called skin carving. That is, the green skin of bamboo is retained for carving with the flesh as the ground color. This is an important craftsmanship. Relief is the shallow carved surface appearing as three dimensional objects. Landscape, figures and animals all may be carved into relief. The relief may be shallow or deep. The deep relief is also called high relief. The shallow convex carving is in fact a kind of shallow relief, with patterns slightly convex. The through carving is the piercing of the bamboo flesh, resulting in a reticulated work. The tondo is a three dimensional object completely carved out of a bamboo material, mostly the root.

The subjects of bamboo sculpture are very extensive. Most of them are landscape, flowers, grasses, trees, birds, figures including those of Buddhism and Taoism. Calligraphy and Chinese painting are the basis of bamboo sculpture. The bamboo craftsmen are always good at writing and painting, or at seal cutting, skillful in artistic planning. Many a refined bamboo work is the result of their excellent expertise. The sculpture of hills, rivers, trees and clouds show their natural pose in twists and turns. Some exquisitely carved flowers and grasses are almost close to the living. Some are fine works of landscape dotted with towers and pavilions, being majestic in all its variety. Some carved figures are lifelike with folds of dress dimly discernible. Some are carved completely out of twisted roots and gnarled branches which are easily executed in cutting and filling. These works own much to consummate skill of craftsmen.

The pictures contained in this album show the style of bamboo sculpture in China. In Ming Dynasty the "Pine and Crane" pen container carved by the great master Zhu Songlin was vivid. On the container an old pine trunk was carved with dense scales and some weird knots. A twisting branch embraced the trunk with exuberant foliage. Under the pine were a pair of cranes, each showing its peculiarity in posture. This work was a birthday gift in honor of some dignity. The name of Zhu Songlin was He, the originator of Jiading school of bamboo sculpture, noted for high relief done by deep cut. He was good at carving pen containers, incense containers, hairpins and goblets. His

方竹笔筒（四川成都）
A brush pot of square bamboo culm (Chengdu, Sichuan)

works were treasured by the then people, but very few were handed down from Ming Dynasty. The "Pine and Crane" pen container was collected by the Nanjing Museum. the "Eight Fairies Pen Container" carries the name of the famous sculptor Pu Zhongqian in Ming Dynasty. On the container there are hanging crags and frowning cliffs, thickly over cast sky and a river with rapid torrent, while on a raft ride eight fairies in great composure. This work is a valuable bamboo sculpture, noted for proper depth and scale of terrain, with objects in picturesque disorder and the whole scene having an imposing air. The master of bamboo sculpture Zhi Yan in Ming Dynasty made a "Landscape Pen Container" which is dark red in color, resulting from skin carving. On the container the landscape shows characteristic smoothness of lines in clear graduation.

In modern times in Kanglu zhuyan (Peaceful House surouded by Bamboo) at Taibei in Taiwan lives a bamboo sculptor Li Youzeng. The antithetical couplets carved by him are noted for nicety of cut and vigorous touches of brush with renovation made on the basis of original characters written. He has won a fame of great bamboo sculptor in Taiwan. In Hunan Province the fine work "Five Hudnred Buddhism Arhats" created by sculptor Tan Nansen. consist of 216 bamboo slips. 1400 cm long, and 54 cm high. with five hundred Buddhism arhats carved on them. by expertise of cutting. It is now in China a large scale bamboo sculpture in a long roll with lifelike figures, noted for coherence and unity. and majesty in representation. In Tongxian Township in Zhejiang Province the "Lotus" and "Calligraphy" and the current fine works of bamboo sculpture in China. The bamboo root sculpture from Xiangshan County in Zhejiang Province is very famous in China. many carving masters are concentrated here. this county was named as " a county of bamboo root carving " in November . 1996 , some of its masterpieces are selected here ."Zhangfei"is also a fine work.

留青笔筒（四川）
A brush pot of green engraving (Sichuan)

Zhangfei's eyebrow, hair and mustache are made of the fibrous roots. His facial expression was heightened, by his beast – like head. glaring eyes, short plump chin, tiger's whiskers and unbearable indignation, as if one could hear his peal – like shouting. Another root sculpture is a " Bust in Muse". The man's configuration shows steeled will, composure and fortitude. The contracted muscles in his face seemed to have elasticity. Bamboo root sculpture "A Childhood pal" expressed the childre's innocence vividly. In Meixian County in Guangdong Province. the bamboo sculpture " Native home of football" is a through carving. showing unique craftsmanship. In Nanjing the bamboo sculpture "Sunzi's Art of War" is a series of bamboo slops, on which 6000 Chinese characters were carved with convex carving technique. The characters carved were graceful and fine. It is a masterpiece of bamboo sculpture.

Playthings made of bamboo branches, tubes and internodes can be seen everywhere at all the tourism sites in South China, catching the fancy of common people. Taking advantage of the natural shapes of bamboo branches and internodes, the craftsmen engaged in sawing and mosaic work created some figures or shapes to fit in with his artistic design. The similarity and dissimilarity of configuration give wide scope to one's imagination. In Nanxiang County in Guangdong Province, the "Bamboo Sailboat", " Leaping Forward" and " Double Deer " full of native colors show full expressiveness. The most fanciful thing is a bamboo coat made of fine tubes taken from the tips of fresh shoots. The surface of the coat is network, giving a handsome novel look and being cool to the flesh of human body.

Figures made of pieces of shoot sheathes glued together are quite vivid. In Leqing County in Zhejiang Province, the figure of " Lord Guangong " made with such a method is impressive both in feature and in manner, just like that described in a popular novel who has a beautiful beard floating before his breast. " Jigong, the legendary monk" and the "God of longevity" are also wonderful to look at.

Besides the bamboo sculptures, carvings, articles made of bamboo branches and tubes, and the figures made of pieces of shoot sheathes, there are other bamboo products such as walking sticks, chopsticks, music instruments, penholders, umbrellas, fans, kites and colored lamps. Please look at the pictures and appreciate the style of variety of bamboo art and crafts.

笔筒（浙江杭州）
A brush pot (Hangzhou, Zhejiang)

寿字扇面（四川）
A covering of fan "Longevity" (Sichuan)

福寿扇面（四川）
A covering of fan "Blessedness and Longevity" (Sichuan)

吉庆如意扇面（四川）
A covering of fan "Auspicious Happiness and Full Satisfaction" (Sichuan)

福富挂件（四川）
"Blessedness and Longevity" (Sichuan)

福字挂帘（四川）
A curtain "Blessedness" (Sichuan)

竹刻书法（湖南衡东，谭南森作）
Engraved calligraphy (Hengdong, Hunan, Tan Nansen)

竹刻书法（湖南衡东，谭南森作）
Engraved calligraphy (Hengdong, Hunan, Tan Nansen)

竹刻长卷《五百罗汉》之一（湖南衡东，谭南森作）
"Five Hundred Arhats", 1 (Hengdong, Hunan, Tan Nansen)

竹刻长卷《五百罗汉》之二（湖南衡东，谭南森作）
"Five Hundred Arhats", 2 (Hengdong, Hunan, Tan Nansen)

竹刻长卷《五百罗汉》之三（湖南衡东，谭南森作）
"Five Hundred Arhats", 3 (Hengdong, Hunan, Tan Nansen)

对联（台湾台北"康庐竹缘"）　Antithetical couplet (Kanglu Bamboo Relations, Taipei, Taiwan)

对联（台湾台北"康庐竹缘"）　Antithetical couplet (Kanglu Bamboo Relations, Taipei, Taiwan)

竹刻长卷《五百罗汉》之四（湖南衡东，谭南森作）
"Five Hundred Arhats", 4 (Hengdong, Hunan, Tan Nansen)

对联（台湾台北"康庐竹缘"）　Antithetical couplet (Kanglu Bamboo Relations, Taipei, Taiwan)

竹刻长卷《五百罗汉》之五（湖南衡东，谭南森作）　"Five Hundred Arhats", 5 (Hengdong, Hunan, Tan Nansen)

作品全长1400厘米，高54厘米，由216块竹简组成，创作者在竹简上镂刻了500个罗汉的形象，组成了一幅形象鲜明、呼应连贯、气势恢宏的艺术长卷，是目前我国最大的竹刻艺术品。
This masterpiece is 1400 cm long and 54 cm wide, consisting of 216 bamboo pieces. The craftsmen created the images of 500 arhats on bamboo, resulting in a vivid, complete and grandiose picture. This is the biggest bamboo engraving article.

对联（台湾台北"康庐竹缘"）　Antithetical couplet (Kanglu Bamboo Relations, Taipei, Taiwan)

臂搁（台湾台北，李佑增作）
Bige (Taipei, Taiwan, Li Youzeng)

线刻笔筒（台湾台北，李佑增作）
An engraved brush pot (Taipei, Taiwan, Li Youzeng)

臂搁仕女图（台湾台北，李佑增作）
(Taipei, Taiwan, Li Youzeng)　Maidens

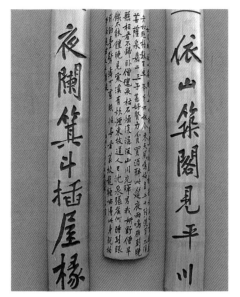

台湾台北"康庐竹缘"拥有600余件竹雕楹联和《兰亭序》、《寒食帖》、《松风阁》等全文竹雕,作者采用了凹雕、凸雕、线雕、嵌雕、浮空雕、穿雕等不同技法,涵盖了楷、行、草、隶、钟鼎、盘铭、甲骨等各种书体,给人一种清新高古的意境美感。

"Bamboo Carving in Kanglu Bamboo Relations" The Kanglu Bamboo Relations Hall in Taipei keeps more than 600 pieces of bamboo carving and engraving, including calligraphy engraving "Foreword to Orchid Pavilion Collection", "Pine Wind Pavilion". Different engraving skills are applied to represent the colorful styles of Chinese calligraphy art, such as regular script, running hand, cursive script, official script, inscription on ancient bronze objects, inscription on bones and tortoise shells, creating aesthetic effects.

对联(台湾台北"康庐竹缘") Antithetical couplet (Kanglu Bamboo Relations, Taipei, Taiwan)

对联(台湾台北"康庐竹缘") Antithetical couplet (Kanglu Bamboo Relations, Taipei, Taiwan)

臂搁书法(台湾台北"康庐竹缘") Calligraphy (Kanglu Bamboo Relations, Taipei, Taiwan)

壁挂小件(台湾台北,李佑增作) A small wall hanging (Taipei, Taiwan, Li Youzeng)

意刻笔筒(台湾台北,李佑增作) An engraved brush pot (Taipei, Taiwan, Li Youzeng)

邮友传谊笔筒(台湾台北,李佑增作) A brush pot of philatelist friend (Taipei, Taiwan, Li Youzeng)

献寿笔筒(台湾台北,李佑增作) A brush pot "Birthday Congratulation" (Taipei, Taiwan, Li Youzeng)

弘一大师笔筒(台湾台北,李佑增作) A brush pot of monk Hongyi (Taipei, Taiwan, Li Youzeng)

竹刻书法（湖南衡东，文国湘、文新学作）
Engraved calligraphy (Hengdong, Hunan, Wen Guoxiang, Wen Xinxue)

留青臂搁"双虾图"（江苏常州，徐秉言作）
Arm pillow of green bamboo "Twin Prawns" (Xu Bingyan, Changzhou, Jiangsu)

竹刻"荷花"（浙江桐乡，叶瑜荪作）"Lotus" (Tongxiang, Zhejiang, Ye Yusun)

《臂搁荷花》
该作品系竹刻陷地深刻，作者以娴熟的刀法，用浮雕法层层深入，表达了荷花、荷叶的体感与质感，在赴新加坡展出时获得好评。

"Lotus"
This bamboo engraving is created with step carving skill, expressing the flowers and leaves distinctively and vividly. It earned favorable comments when displayed in Singapore.

留青臂搁书法（江苏常州，徐云作）
Arm pillow of green bamboo with calligraphy (Xu Yun, Changzhou, Jiangsu)

竹刻书法（浙江桐乡，叶瑜荪作）Engraved calligraphy (Tongxiang, Zhejiang, Ye Yusun)

钟馗（浙江象山）Deity Zhong Kui chasing demons (Xiangshan, Zhejiang)

足球之乡（广东梅县）A country of soccer (Meixian, Guangdong)

竹刻书法长卷"千禧万寿百龙图"之一（湖南衡东，谭南森作）
Engraved bamboo calligraphy "Numerous blessedness, longevity and dragons" I (Tan Nansen, Hengdong, Hunan)

竹刻书法长卷"千禧万寿百龙图"之二（湖南衡东，谭南森作）
Engraved bamboo calligraphy "Numerous blessedness, longevity and dragons" II (Tan Nansen, Hengdong, Hunan)

留青臂搁"新篁"（浙江桐乡，叶瑜荪作）Arm pillow of green bamboo "New Bamboo" (Ye Yusun, Tongxiang, Zhejiang)

留青臂搁书法（浙江桐乡，叶瑜荪作）Arm pillow of green bamboo with calligraphy (Ye Yusun, Tongxiang, Zhejiang)

阴刻臂搁书法（浙江桐乡，叶瑜荪作）Engraved arm pillow with calligraphy (Ye Yusun, Tongxiang, Zhejiang)

浮雕臂搁"云龙图"（浙江桐乡，叶瑜荪作）Engraved arm pillow "Dragon in Clouds" (Ye Yusun, Tongxiang, Zhejiang)

留青、阴刻"踏雪寻梅图"臂搁（浙江桐乡，叶瑜荪作）Engraved arm pillow of green bamboo "Looking for Plum in Snow" (Ye Yusun, Tongxiang, Zhejiang)

阴刻沙底"吴昌硕'寿'字"臂搁（浙江桐乡，叶瑜荪作）Engraved arm pillow with emblem of longevity (Ye Yusun, Tongxiang, Zhejiang)

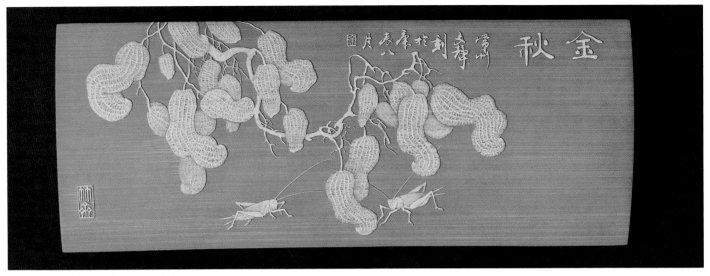

留青臂搁"金秋"（江苏常州，徐文静作）
Arm pillow of green bamboo "Golden Autumn" (Xu Wenjing, Changzhou, Jiangsu)

留青臂搁"青白乐"（江苏常州，陆颖毫作）
Arm pillow of green bamboo "Green vegetables and white radishes" (Lu Yinghao, Changzhou, Jiangsu)

翻簧笔筒（浙江黄岩，罗启松作）
Brush pot (Luo Qisong, Huangyan, Zhejiang)

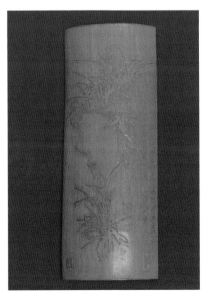

留青臂搁"路回兰"（浙江桐乡，钟山隐作）
Arm pillow of green bamboo "Magnolia on Returning Path" (Zhong Shanyin, Tongxiang, Zhejiang)

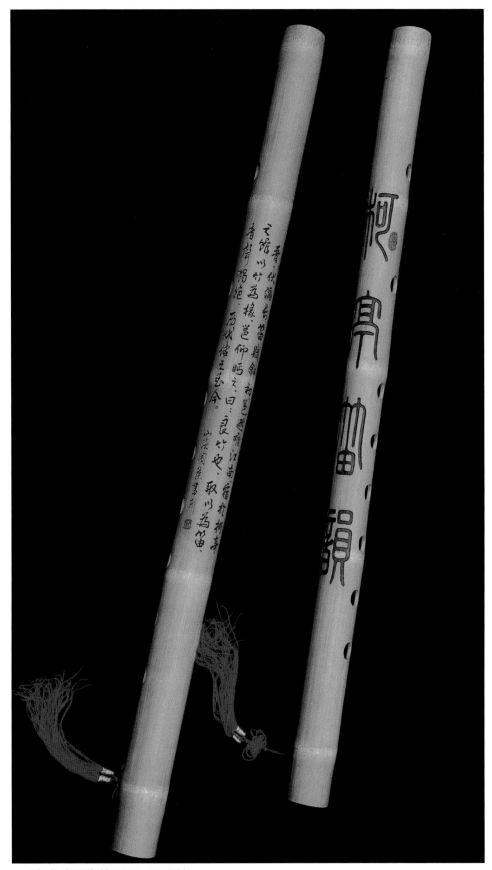

竹刻"柯亭笛韵"（浙江绍兴，王国荣作）
Engrave bamboo "Flute Sound from Pavilion" (Wang Guorong. Shaoxing, Zhejiang)

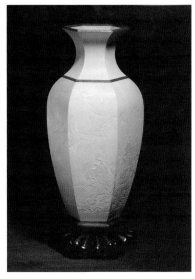

翻簧"松鹤花瓶"（浙江黄岩，罗启松作）
A Vase decorated with the figures of pine and crane (Luo Qisong, Huangyan, Zhejiang)

翻簧竹扇线刻仕女图（浙江黄岩，罗启松作）
Bamboo fan decorated with lady figures (Luo Qisong, Huangyan, Zhejiang)

翻簧掌扇（浙江黄岩，罗启松作）
Bamboo fan (Luo Qisong, Huangyan, Zhejiang)

扇形竹刻书法（湖南衡东，谭南森作）
Calligraphy in fan-shape (Tan Nansen, Hengdong, Hunan)

留青茶叶筒（江苏常州，徐素白作）
Green tea tube of bamboo (Xu Subai, Changzhou, Jiangsu)

留青臂搁"孔雀"（江苏常州，徐秉言作）
Arm pillow of green bamboo "Peacock" (Xu Bingyan, Changzhou, Jiangsu)

竹刻"梅、兰、竹、菊"书画条屏（湖南衡东，谭南森作）
Egraved bamboo screen "Plum, Magnolia, Bamboo and Chrysanthemum" (Tan Nansen, Hengdong, Hunan)

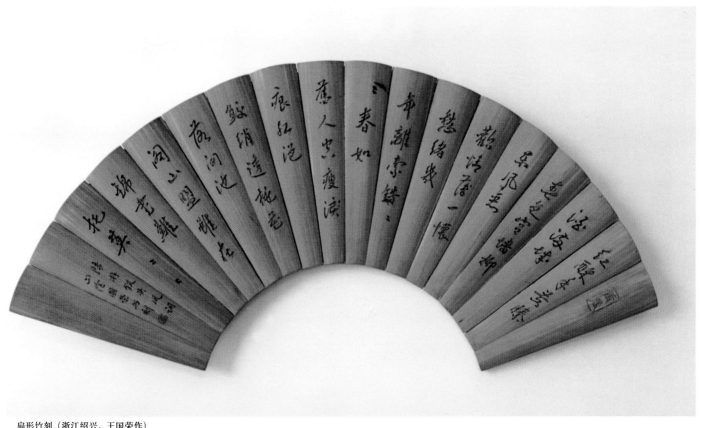

扇形竹刻（浙江绍兴，王国荣作）
Fan-shaped bamboo carving (Wang Guorong, Shaoxing, Zhejiang)

留青臂搁"荷"（江苏常州，徐秉言作）
Arm pillow of green bamboo "Lotus" (Xu Bingyan, Changzhou, Jiangsu)

留青臂搁"秋之歌"（江苏常州，徐秉言作）
Arm pillow of green bamboo "Autumn Song" (Xu Bingyan, Changzhou, Jiangsu)

留青臂搁"兰花"（江苏常州，沈竹舟作）
Arm pillow of green bamboo "Magnolia" (Shen Zhuzhou, Changzhou, Jiangsu)

天女散花笔筒（台湾台北，李佑增作） A brush pot "Heavenly Maiden Scattering Flowers" (Taipei, Taiwan, Li Youzeng)

张飞（浙江象山，张德和作）
Zhang Fei (Xiangshan, Zhejiang, Zhang Dehe)

济公（浙江象山，张德和作）
Monk Jigong (Xiangshan, Zhejiang, Zhang Dehe)

沉思（浙江象山） Pondering (Xiangshan, Zhejiang)
该作品仿照美国总统林肯像进行施艺，作者对竹根须作了艺术的处理，使其巧妙地成为头像的须发，作品高29厘米，宽18厘米，是"中国竹根雕艺术之乡"的绝品。
This article was created to reproduce the portrait of Abraham Lincoln, the former US president. Cleverly turning the bamboo root hair into the hair of head in the portrait. This masterpiece is 29 cm in height, 10 cm in width. This is a top grade from the "Country of Bamboo Weaving".

王昭君（浙江象山，张德和作）　Wang Zhaojun, a beautiful maiden in Han Dynasty (Xiangshan, Zhejiang, Zhang Dehe）一段 50 厘米长的竹根，在艺人的刀下，竟化为一个楚楚动人的绝代美女王昭君。作品充分发挥竹质的韵味；根蒂和须毛处理成发髻和饰品，须毛节间被处理成毛茸茸的大氅，极富装饰味和韵律感。获中国根艺优秀作品展《刘开渠根艺奖》金奖。
A piece of bamboo root 50 cm long, through the magic knife of the craftsman, turned into Wang Zhaojun, a woman of matchless beauty, delicate and charming. The root hair is utilized to reproduce the furcoat and ornamentals of the beauty. It has won the first prize of "Liu Kaiqu Root Carving Art".

寿星（浙江象山，张德和作）
God of Longevity (Xiangshan, Zhejiang, Zhang Dehe)

面具（四川）
Masks (Sichuan)

苏武（浙江象山，张德和作）　Su Wu, an envoy of Chinese emperor in Han Dynasty (Xiangshan, Zhejiang, Zhang Dehe)

蓑笠翁（浙江宁波，杨古城作） An old man in bamboo cap (Ningpo, Zhejiang, Yang Gucheng)
该作品体现了"孤舟蓑笠翁，独钓寒江雪"的诗意，在1990年第九届中国工艺美术品百花奖评比会上荣获创新设计一等奖。
This article reproduces the poetry of "An old fishing man in bamboo cap, from a solitary small boat, on the snow-covered river". It won the first-class prize of new design on the Chinese Handicrafts Hundred Flowers Competition in 1990.

子孙万代（浙江象山，张德和作） Blessed with many children and posterity (Xiangshan, Zhejiang, Zhang Dehe)

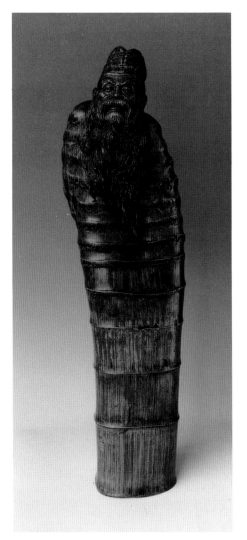

遐思（浙江象山，张德和作）
Reverie (Xiangshan, Zhejiang, Zhang Dehe)

面具（四川） A mask (Sichuan)

三脚鼎与手提罐（浙江象山）
Three legged incense burner and a hand-pot (Xiangshan, Zhejiang)

竹雕水筒（浙江象山）
Bamboo water pots (Xiangshan, Zhejiang)

钟馗（浙江象山，张德和作）
Zhong Kui, a deity chasing demons (Xiangshan, Zhejiang, Zhang Dehe)

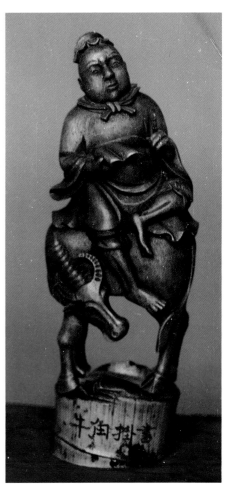

牛角挂书（湖南南岳，梁丰助作）
Books hanged on ox horn (Nanyue, Hunan, Liang Fengzhu)

"鬼谷子"（浙江象山，陈善国）
"Wise ghost" (Chen Shanguo, Xiangshan, Zhejiang)
巧用一段畸形竹根，活脱脱地刻画了鬼谷子的形象，荣获浙江省首届雕刻型根艺大奖展"金奖"。
A piece of abnormal bamboo root was utilized efficiently, the figure of a wise ghost was described vividly, winner of a gold prize at Zhejiang Fair of Root-carving.

赶潮去（浙江宁海，葛伟峰作）
To look at the waves (Ge Weifeng, Ninghai, Zhejiang)

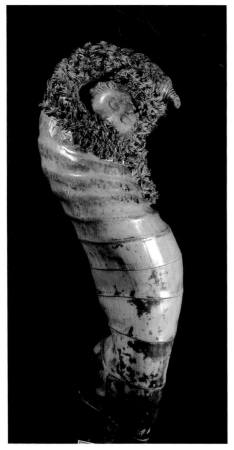

早春如夜（浙江象山，王群作）
Early spring as at night (Wang Qun, Xiangshan, Zhejiang)

松鹤笔筒（明 朱松邻）A brush pot "Pine and Crane" (Zhu Songlin, Ming Dynasty)
取材于一段老竹，刻成一截虬枝纷批、曲折盘旋的老松，松针重叠，层次分明，松间一对仙鹤，顾盼有情，使静谧的松林增添了勃勃生机。笔筒高17.8厘米，直径8.9～14.9厘米，是一件难得的精品，现珍藏于南京博物馆。
A piece of old bamboo tube has been used to reproduce an old pine, with a crooked trunk and numerous branches. Two cranes linger under the pine, gazing on each other, which vitalizes the quietness of pine forest. This article is 17.8 cm high and 8.9 to 14.9 cm in diameter. This is an extraordinary masterpiece.

舞姿（浙江象山，张松林作）
Dancing figure (Zhang Songlin, Xiangshan, Zhejiang)

冲浪（浙江宁海，葛伟峰作）
Over the waves (Ge Weifeng, Ninghai, Zhejiang)

灵山归来（浙江象山，朱利勇作）
Returning from Lingshan (Zhu Liyong, Xiangshan, Zhejiang)

八仙笔筒（明　濮仲谦）A brush pot "Eight Immortals" (Pu Zhongqian, Ming Dynasty)

山水纹笔筒（明　芷岩）A brush pot (Zhi Yan, Ming Dynasty)

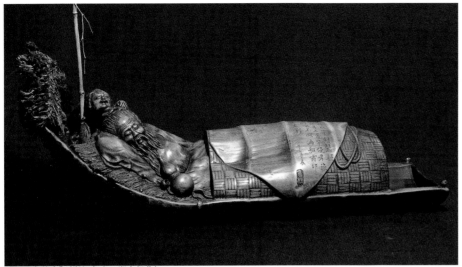

"相聚在梦中"（浙江象山，郑宝根作）
"Gathering in dream" (Zheng Baogen, Xiangshan, Zhejiang)
作品雕琢精致，意境深远，促人联想。荣获浙江省首届雕刻型根艺大奖展"金奖"。 Carved delicately, with a vivid artistic conception, winner of gold prize at Zhejiang Fair of Root-carving.

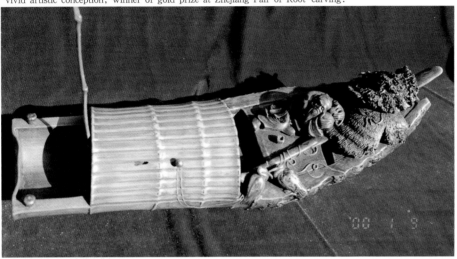

酣（浙江象山，张德和作）
Sleeping deeply (Zhang Dehe, Xiangshan, Zhejiang)
获中国第七届根艺美术作品展金奖。Winner of a gold prize at 7th Fair of Chinese Root-carving

家家扶得醉人归（浙江象山，张德和作）
a drunken lady leaning on a maid (Zhang Dehe, Xiangshan, Zhejiang)

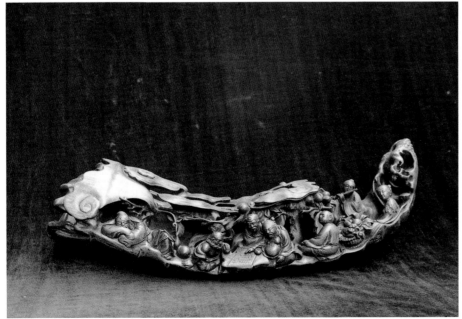

八仙过海（浙江象山，郑宝根作）
Eight deities over the sea (Zheng Baogen, Xiangshan, Zhejiang)

母爱（浙江宁海，葛伟峰作）
Mother love (Ge Weifeng, Ninghai, Zhejiang)

江边蓑笠翁（浙江宁海，葛安飞作）An old fisherman on riverside (Ge Anfei, Ninghai, Zhejiang)
巧用材质，雕功娴熟，是一件艺术品位高的佳作。A piece of bamboo material processed efficiently, a masterpiece of root-carving.

踏雪折梅（浙江象山，何益平作）
To get plum flowers in snow (He Yiping, Xiangshan, Zhejiang)

温暖人间（浙江象山）
The world warmed up (Xiangshan, Zhejiang)

渔舟晚唱（浙江象山，张德和作）
Evening songs from fishing boat (Zhang Dehe, Xiangshan, Zhejiang)

窥视人间（浙江象山，郑宝根、徐亚玲作）
A glance at the human world (Zheng Baogen, Xu yaling, Xiangshan, Zhejiang)
获中国第七届根艺美术作品展金奖。Winner of a gold prize at 7th Fair of Chinese Root-carving.

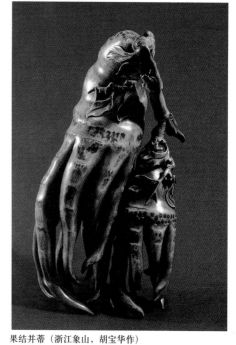

果结并蒂（浙江象山，胡宝华作）
A pair of fruits (Hu Baohua, Xiangshan, Zhejiang)

刘海喜庆（浙江象山，陈善国作）
A holiday of Liuhai (Chen Shanguo, Xiangshan, Zhejiang)

童年的回忆（浙江象山，蔡海楚作）
Recollecting childhood (Cai Haichu, Xiangshan, Zhejiang)

布袋和尚（浙江象山，张德和作）
A monk with a textile bag (Zhang Dehe, Xiangshan, Zhejiang)

面对人间不平事（浙江象山，周秉益作）
Facing unfairness in the world (Zhou Bingyi, Xiangshan, Zhejiang)

老寿星（浙江象山）
The god of longevity (Xiangshan, Zhejiang)

春山独钓翁（浙江象山，陈春望作）
A old man fishing alone (Chen Chunwang, Xiangshan, Zhejiang)

双凤宝瓶（浙江象山，周秉益作）
Twin phoenix (Zhou Bingyi, Xiangshan, Zhejiang)

笔筒"赏梅"（浙江象山，钱锁挺作）
Brush pot "Enjoying Plum" (Qian Suoting, Xiangshan, Zhejiang)

竹根造型"春色满园"（浙江嵊州，徐祖望、商华水作）
Spring garden (Xu Zuwang, Shang Huashui, Shengzhou, Zhejiang)
作者巧用凤尾竹竹根，创作出此件罕见作品，在浙江省首届雕刻型根艺大奖展中荣获"金奖"。A infrequency masterpiece created artfully from phoenix bamboo, winner of gold prize at the First Zhejiang Fair of Root-carving.

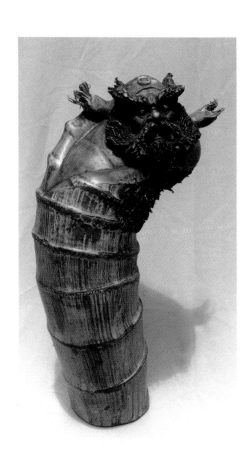

"怒见不平一声吼"（浙江嵊州，郑兴国作）
"Roar against discontentment" (Zheng Xingguo, Shengzhou, Zhejiang)

笔筒"松荫雅集"（浙江象山，钱锁挺作）
Brush pot "Gathering under Pines" (Qian Suoting, Xiangshan, Zhejiang)

喜娃（浙江象山，朱利勇作）
Baby Xiwa(Zhu Liyong, Xiangshan, Zhejiang)

大山里的竹农（浙江象山，朱利勇作）
A Bamboo farmer in mountain area (Zhu Liyong, Xiangshan, Zhejiang)

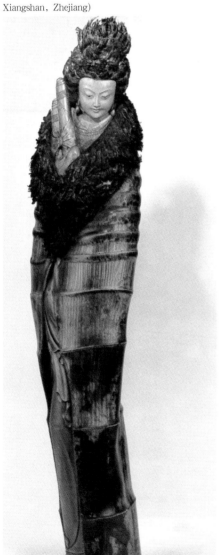

仕女（浙江象山，张德和作）A beautiful lady (Zhang Dehe, Xiangshan, Zhejiang)
获1999浙江根艺精品大展金奖。
Winner of a gold prize at Zhejiang Fair of Root-carving in 1999.

虚怀劲节（浙江象山，张德和作）
Modesty and chastity (Zhang Dehe, Xiangshan, Zhejiang)

早春寻梦（浙江象山，王群作）
Seeking sweet dream in early spring (Wang Qun, Xiangshan, Zhejiang)
作品刻画了一位纯净的江南小镇少女形象，富有浓郁的诗意与乡土气息。在浙江省首届雕刻型根艺大奖展上荣获"金奖"。A little girl from a small town in southern area, full of rural smells, winner of gold prize at the First Zhejiang Fair of Root-carving.

大禹治水（浙江象山，张德和作）
Dayu the Great controls water flood (Zhang Dehe, Xiangshan, Zhejiang)

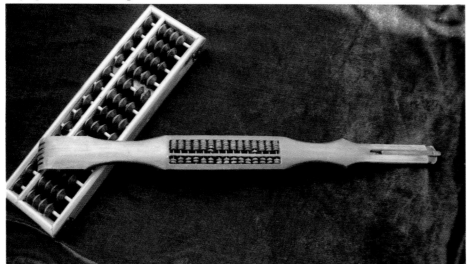

算盘、抓痒巴与挖耳勺（福建泉州）
Abacus, itch-scratching hook and earpick (Quanzhou, Fujian)

谆谆教诲（浙江象山，张先学作）
Teaching wholeheartedly (Zhang Xianxue, Xiangshan, Zhejiang)

寒岁三友（浙江宁海，葛安飞作）
Three friends in winter (Ge Anfei, Ninghai, Zhejiang)

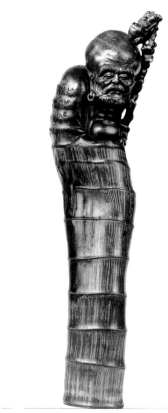

持杖罗汉（浙江象山，张德和作）A Buddha disciple with a stick (Zhang Dehe, Xiangshan, Zhejiang)

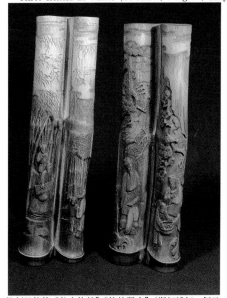

竹刻双笔筒"仙人竹鹤""竹林琴声"（浙江浦江，何云青作）Twin brush pots "Celestial being, bamboo and crane", "Music from Bamboo Grove" (He Yunqing, Pujiang, Zhejiang)

又是一个丰年（浙江象山，周秉益作）
An abundant year again (Zhou Bingyi, Xiangshan, Zhejiang)

闲趣（浙江象山，张德和作）Enjoying a rest (Zhang Dehe, Xiangshan, Zhejiang)
获中国第五届根艺优秀作品展"刘开渠根艺奖"银奖。
Winner of a silver prize at 5th Fair of Chinese Root-carving.

"盼"（浙江象山，陈善国、应芬芬作）
"Waiting" (Chen Shanguo, Ying Fenfen, Xiangshan, Zhejiang)

作品刻画了一位年迈父亲期盼儿子平安回来的情景。
An old father waiting for his son.

远瞩（浙江嵊州，徐华铛）
Looking far forth (Xu Huadang, Shengzhou, Zhejiang)

探道（浙江象山，张德和作）
Path exploring (Zhang Dehe, Xiangshan, Zhejiang)

小伙伴（浙江象山，郑宝根作） Young companions (Zheng Baogen, Xiangshan, Zhejiang)

警觉（浙江象山，周秉益作）
Watchfulness (Zhou Bingyi, Xiangshan, Zhejiang)

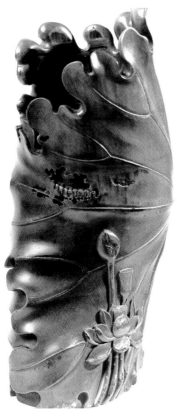

残荷归风（浙江象山，张德和作）
Remnant lotus in wind (Zhang Dehe, Xiangshan, Zhejiang)

浮雕插笔筒（浙江象山，张德和作）
Engraved brush pot (Zhang Dehe, Xiangshan, Zhejiang)

高原情（浙江象山，郑宝根作）
A smiling girl form plateau (Zheng Baogen, Xiangshan, Zhejiang)

孔雀（四川大足）
Pearl Harbor (Dazu, Sichuan)

智者（浙江象山，张德和作）
A wise man (Zhang Dehe, Xiangshan, Zhejiang)

推敲（浙江象山，张德和作）
Considering (Zhang Dehe, Xiangshan, Zhejiang)

竹片画（浙江义乌）
Picture of bamboo strips (Yiwu, Zhejiang)

张飞与赵云（浙江乐清，吴涛林作）
Two heroes, Zhang Fei and ZhaoYun (Wu Taolin, Leqing, Zhejiang)

乐器——空竹（浙江安吉，中国竹子博物馆）
Musical instrument—empty bamboo (Chinese Bamboo Museum, Anji, Zhejiang)

"除恶图"仿任伯年笔意（浙江嵊州，赵国华烙）
Picture "Eradicating evils", imitated from the painting of Ren Bonian (Zhao Guohua, Shengzhou, Zhejiang)

乐器——箜篌与德朗琴（浙江安吉，中国竹子博物馆）
Musical instrument—"Konhou" and "Delang" (Chinese Bamboo Museum, Anji, Zhejiang)

民族竹用具 Bamboo tools of national minorities

傣家小竹箩"扁帕",是傣族群众上山打猎和下地劳动必然配带的工具。
A small bamboo basket of Dai people, which is generally bring in hunt and work in farm.

傣家竹饭盒。
A bamboo lunch box of Dai people.

农村集市上的传统竹制品。
The bamboo goods in the market of rural area.

云南边疆各族人民对竹子利用历史悠久,竹子与各族人民的生产、生活、历史、文化、艺术、宗教、习俗、以及衣、食、住、行、用息息相关,形成了独特的民族竹文化,被誉为"民族竹文化之乡"。瑰丽多姿的少数民族竹文化是云南珍贵民族文化的典型代表之一。(摄影 辉朝茂、薛嘉榕)

In the long history of production, people of various nationalities in Yunnan have been using bamboos in many aspects of their life such as clothes, food, transportation and dwelling. Bamboo and the production and living of the people are closely related. Bamboo culture is a representative of the valuable culture of ethnic groups from Yunnan, because it has deep and special influences in Yunnan people's history, culture, drawing, music, religion, custom, building, and agriculture production. (Photos: Hui Chaomao & Xue Jiarong)

新平花腰傣族的小竹箩。
A beautiful bamboo basket of Huayao Dai people which living in Xinping county.

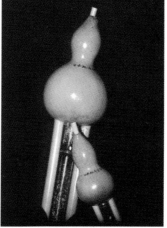

传统的云南民族竹乐器葫芦丝,因其音质纯朴、音色优美而享誉中外。
The traditional musical instrument in Yunnan, which made in gourd and bamboo.

陇川景颇族竹筒茶,获得第三届中国竹文化节金奖。
The product of tea which packed by bamboo canister.

竹草鞋伴随山区少数民族从历史走来。
Bamboo sandal is widely uaed in the rural area.

傣家竹茶几。The tea and dining table made in bamboo.

传统渔具离不开竹编。
Many of the fishing tackles are made in bamboo.

竹制小工艺 Small bamboo handicrafts

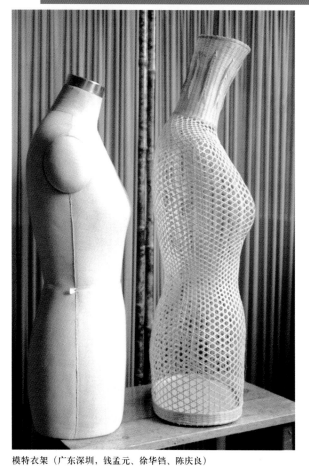

一片竹制成的果篮（浙江杉华，史舟棠）
A fruit basket made of a single piece of bamboo material (Shi Zhoutang, Shahua, Zhejiang)

模特衣架（广东深圳，钱孟元、徐华铛、陈庆良）
Clothes Stands (Qian Mengyuan, Xu Huadang, Chen Qingliang, Shenzhen, Guangdong)

蟹（四川大足）Crab (Dazu, Sichuan)

篮胎漆花瓶（浙江嵊州）
A vase with double handles (Shengzhou, Zhejiang)

荷花篮（浙江象山）
Lotus basket (Xiangshan, Zhejiang)

贴花蝶形花瓶 Betterfly-shaped vases

龙船与官船（广东）A dragon boat and amandarin's boat (Guangdong)

头像（四川成都）Heads (Cengdu, Sichuan)

双鹿(四川渠县)
Two deers (Quxian, Sichuan)

飞跃(广东南雄,陈继彭作)
Jumping (Nanxiong, Guangdong, Chen Jipeng)

玩具小车
Toy carts

玩具躺椅、屏风
Toy deck chairs, screen

玩具小车、小船
A toy cart, boat

竹草人物玩具
Toy figures and playthings

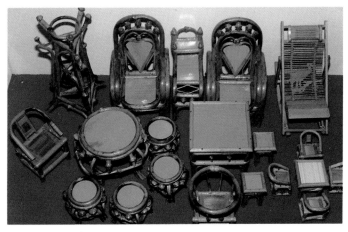
竹制工艺品(福建顺昌)　Bamboo handicrafts (Shunchang, Fujian)

竹制工艺品(福建顺昌)　Bamboo handicrafts (Shunchang, Fujian)

帆船(福建顺昌)　Sailing boat (Shunchang, Fujian)

蝴蝶挂扇（竹片造型）（四川）
Butterfly fans (Sichuan)

雄鹰（笋壳贴画）（四川）
A valiant eagle (Sichuan)

鸟笼
A bird cage

济公（笋壳贴）（浙江温州，吴涛林作）
Monk Jigong (Wenzhou, Zhejiang, Wu Taolin)

麻鸭（笋壳贴）（浙江）
Ducks (Zhejiang)

老寿星（笋壳贴）（浙江温州，吴涛林作）
Longevity (Wenzhou, Zhejiang, Wu Taolin)

鹰（笋壳贴）（广西）
An eagle (Guangxi)

翠竹情（笋壳贴画）（四川）
Enjoying green bamboo (Sichuan)

鸟笼（广东陆丰）
Bird cages (Lufeng, Guangdong)

山水画帘（四川）
A landscape curtain (Sichuan)

松鹤画帘（四川）
A curtain "Pine and Crane" (Sichuan)

鸟窝（浙江嵊州）
Bird nests (Chengzhou, Zhejiang)

糖果盒（翻黄）（四川）
A candy pot (Sichuan)

熊猫画帘（四川）
A curtain "Panda" (Sichuan)

竹制旅行箱（浙江嵊州）
Bamboo suitcase (Chengzhou, Zhejiang)

竹制旅行箱内部（浙江嵊州） Bamboo suitcase, inside view (Chengzhou, Zhejiang)

日式竹茶具（四川成都） Bamboo tea set of Japanese style (Chengdu, Sichuan)

筷筒（福建泉州）
Chopstick holders (Quanzhou, Fujian)

室内挂饰之一（上海，胡曰龙作）
Interior hanging, 1 (Shanghai, Hu Yuelong)

傣族花伞
Flower umbrella of Dai tribe

芦笙
A reed pipe wind instrument

竹片造型（湖南）
Bamboo models (Hunan)

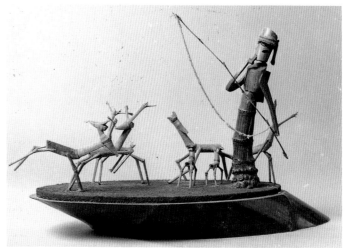
春之歌（广东南雄，陈继彭作）
A song of spring (Nanxiong, Guangdong, Chen Jipeng)

茶叶罐（翻黄）（四川）
A tea pot (Sichuan)

竹骨绍兴王星记折扇珍品（浙江绍兴）
An excellent fan (Shaoxing, Zhejiang)

梳篦（江苏）
Thick-and Fine-toothed combs (Jiangsu)

THE WORLD OF BAMBOO MATERIAL

竹家具 Bamboo furniture

竹桌、竹椅（四川成都） Bamboo table and chairs (Chengdu, Sichuan)

竹子除用于劈篾编织和雕刻成器外，还广泛用于家具、建筑、交通运输和其它部门。

竹器家具制作轻巧、典雅、隽秀、价格低廉，在中国南方诸省（区）使用非常普遍。一般选用材质坚韧的毛竹、刚竹、桂竹、茶杆竹、刺竹、紫竹、水竹等竹种，对竹杆进行卷节、火烤、弯曲、薰蒸、打穴、凿孔、开糟、榫合、竹条或竹片拼面等工序后，制成桌、椅、床、橱、茶几、花架等多种家具。近年来，运用先进的加工技术，将竹子经过水煮、刨削、干燥、弯曲成型、胶合、砂光等多种工序，制成直线型、曲线型的柱材及零件、板材。再按照现代家具的制作工艺，加工成各种造型别致、防虫柱、防开裂、榫合紧密、坚固耐用的各型竹家具。

竹建筑多见于亭、台、楼、阁、水榭等。用竹制作建筑材料，在中国古籍中不乏记载。清《粤西琐记》便有"不瓦而盖，盖以竹；不墙而墙，墙以竹；不板而门，门以竹，其余若楞、若椽若窗、若承壁，莫非竹者，吾署上房，亦竹屋"的记载。中国南方属热带、亚热带地区，炎热多雨，人们多搭建竹楼，清凉避湿。在西双版纳，更有风格独特的傣家竹楼，楼顶像诸葛亮的帽子，一座挨着一座。竹楼的墙用竹片排列编成，楼板用竹器镶接。中国的风景园林点也往往以竹楼点缀其间，供游人休闲。杭州西湖水榭式长廊，其顶、梁、柱、檩、栏杆等全用竹子制成，和湖光山色相映成趣。南京玄武湖一排竹亭立于水面之上，入夜灯光溢彩，别具情趣。各地宾馆、饭店、酒家、餐厅采用竹子装潢，都给人一种清新典雅、返朴归真的享受。

中国南方有一种用毛竹排列穿缀制成的竹筏，浮在水面上，既可载货也可载客，这种传统的运输工具现已成为旅游风景点的一种游览工具，旅客乘竹筏顺流而下，观赏风景，奇山异石尽收眼底，身临其境，心旷神怡。

竹子与木材相比，它的直径小、壁薄中空、尖削度大，故现有的木材加工工艺和设备在竹材加工中都不能直接使用。科技人员经过多年的刻苦攻关，采用竹片胶合、竹篾胶合、竹材碎料胶合等现代加工综合方法，对竹材进行工业化利用，研制开发了竹材胶合板系列产品。在汽车、建筑、集装箱等工业部门及人们的生活中得到了广泛的利用。如用竹材胶合板作载重汽车、客车的车厢底板，不仅牢固坚实，而且经济实惠；高强覆膜竹胶合模板、竹席竹帘复合胶合模板，在我国建筑工业上应用后，受到普遍的欢迎；竹拼花地板以它典雅耐磨的特色，源源不断地进入寻常百姓人家；竹木复合集装箱底板已成功地取代硬木胶合底板，成了集装箱制造中的一支生机勃勃的生力军；以竹材采伐和加工的剩余物和小径竹为原料的竹材碎料，也化废为宝，制成竹材碎料板，广泛应用于建筑模板和包装材料中。

中国的竹材利用虽还年轻，但潜力深厚，方兴未艾，勤劳智慧的中国科技人员与能工巧匠们，正以自己的心血和汗水把根根翠竹化解成多姿多彩的竹材世界。

圆桌、圆凳（四川成都）
Round table and round stools (Chengdu, Sichuan)

Bamboo culms may be cut into slits used in weaving. Some articles are completely carved out of bamboo or made of bamboo tubes and slips on wihich patterns are engraved or carved. Banboo is of wide use. It may be used in furniture making, construction, decoration, communication, transportation and other industries.

Bamboo furniture is light, delicate, handsome and less expensive. It is widely used in South China. The firm and tough bamboo materials chosen are usually those of *Phyllostachys heterocycla* var. *pubescens*, *Ph. sulfurea* var. *viridis*, *Ph. bambusoides*, *Pseudosasa amabilis*, *Chimnobambusa pachystachys*, *Phyllostachys nigra*, and *Ph. heteroclada*. In processing the joints of culms are removed, the culms are roasted over fire, bent, fumigated, punctured, notched and tenoned. Bamboo slits and slops are matched furniture are made, such as desks, chairs, beds, cupboards, tea tables, and flower pot supports. In recent years masters of furniture making applied bamboo material to replace wood and created bamboo furniture better than wooden. bamboos are boiled, bent, molded, glued and polished with abrasive paper with sophisticated methods to be made into linear or curved pillars, parts and boards. Then in accordance with modern manufacturing process they are made into various sorts of fashionable furniture which are borer proof, crack proof and tightly mortised durables. The common bamboo buildings are pavilions, platrforms, stories, towers, and waterside pavilions.

Records of bamboo used as construction material can frequently be found in ancient Chinese classics. In Qing Dynasty there was a statement in "Trivial Records of West Guangdong": The roof is not covered with tile, but with bamboo, the wall is not made of bricks and earth, but with bamboo tubes. Other parts of structure such as arris, rafter, window and partition wall are all bamboos. The main rooms in may office are also built with bamboos". South China belongs to tropical and subtropical zone, where it is very hot and with plenty of rain. Bamboo stories are built to avoid heat and to enjoy cool air. In Xishuangbanna in Yunnan Province bamboo towers with a particular style are built by the Tai race. The roof of the towers standing closely near each other is like the Zhuge Liang's (Zhuge Liang was a famous statesman in the period of Three Kingdoms) steep sloping hat-top. The wall is made of bamboo slips well weaved or arranged. The floor is made of parqueted bamboo tubes. The landscape gardens in China are often dotted with bamboo towers, in which the sightseers may rest. The roofs, beams, pillars, eaves and rails of the pavilion-like corridor in West Lake are all made of bamboos.

By side of the corridor the glimmering lake and the misty hill appear pleasing and enchanting. In Xuanwu Lake in Nanjing, several rows of bamboo pavilions stand above water surface. In night the colored lights make rainbow, being particularly interesting. The bamboo decorations in hotels, restaurants, wine shops and dining rooms in various localities are fresh and elegant, where one can enjoy the true charms of nature.

In South China there is a kind of raft made of moso bamboo flowing on river, which may carry both cargo and passengers. This traditional water transporting equipment now has become a recreational tool in scenic spots.

The sightseers riding this raft flowing down stream or river may enjoy the scenery. When you are here with wonderful hills and rocks in sight, you will feel carefree and joyful.

Compared to wood, the bamboo being of small diameter have hollow culms with thin walls, and sharp tapering shoot. The now available processing technology and equipment for wood cannot be directly applied to the processing of bamboos. The Chinese technicians after many years' painful effort in problem attacking have developed some modern integrated processing methods for bamboo, such as gluing together the bamboo slips, skin and particles, industrializing the use of bamboo and developing a series of plybamboo, widely used in automobile manufacture, construction, container making industry as well as in daily life. Batches and batches of parquet bamboo floors characterized by wear resistance and elegant style flow into common households. The hardwood plywood bottom boards of containers have been successfully replaced by the bamboo-wood composite boards.

This board has become a necessary and important material in container making industry in China. The slash after bamboo felling and small-diameter bamboos being all wastes have turned into valuable materials. The bamboo particle boards made of these materials have been widely used in manufacture of box pattern plates in construction and used as packing materials in various industries.

The bamboo industry in China is still in infancy, but it has tremendous potentialities and boosts an increasing development. The clever and diligent Chinese technicians and the skillful craftsmen are engaged in the work with their energy and sweat to turn the green bamboos into a variegated and splendid bamboo world.

方桌、方凳（四川成都） Square table and square stools (Chengdu, Sichuan)

小方桌、小方椅（江苏南京） Small square table and small square stools (Nanjing, Jiangsu)

竹椅茶几（四川成都）
Bamboo chair and tea table (Chengdu, Sichuan)

竹椅茶几（四川成都）
Bamboo chairs and tea table (Chengdu, Sichuan)

躺椅（江苏宜兴）
Deck chairs (Yixing, Jiangsu)

躺椅（四川成都）
A deck chair (Chengdu, Sichuan)

长桌（江苏南京）
A long table (Nanjing, Jiangsu)

竹制沙发套装（浙江新昌，南洲竹业）
Bamboo sofa (Nanzhou Bamboo Processing Enterprise, Xinchang, Zhejiang)

鼓桌凳（浙江安吉，曾伟人作）
Table and stools of bamboo (Zeng Weiren, Anji, Zhejiang)

逍遥椅（浙江安吉，曾伟人作）
Rocking chair (Zeng Weiren, Anji, Zhejiang)

竹制坐椅（浙江新昌，南洲竹业）
Bamboo armchair (Nanzhou Bamboo Processing Enterprise, Xinchang, Zhejiang)

竹制麻将桌椅（浙江新昌，南洲竹业）
Bamboo card-playing table and chairs (Nanzhou Bamboo Processing Enterprise, Xinchang, Zhejiang)

竹制圆桌（浙江新昌，南洲竹业）
Bamboo round table (Nanzhou Bamboo Processing Enterprise, Xinchang, Zhejiang)

竹制餐桌椅（浙江新昌，南洲竹业）
Bamboo dining table and chairs (Nanzhou Bamboo Processing Enterprise, Xinchang, Zhejiang)

花竹沙发套装（浙江安吉，曾伟人作）
Armchairs of patterned bamboo (Zeng Weiren, Anji, Zhejiang)

花竹椅几（浙江安吉，曾伟人作）
Chairs and teatable of patterned bamboo (Zeng Weiren, Anji, Zhejiang)

竹制花架（浙江安吉，曾伟人作）
Flower rack of Bamboo (Zeng Weiren, Anji, Zhejiang)

橱（江苏南京）
A cabinet (Nanjing, Jiangsu)

小方桌（江苏南京）
A small square stool (Nanjing, Jiangsu)

橱（江苏南京）
A cabinet (Nanjing, Jiangsu)

小茶几（江苏南京）
A small tea table (Nanjing, Jiangsu)

蓝胎漆圆桌圆凳与屏风
Lacquered round table, round stools and screen

六角小桌（江苏南京）
A small hexagon table (Nanjing, Jiangsu)

竹屏风（浙江东阳）
Bamboo Screen (Dongyang, Zhejiang)

竹屏风（浙江东阳）
Bamboo Screen (Dongyang, Zhejiang)

竹装潢 Bamboo decoration

东阳市招待所竹厅之二（浙江东阳）
Bamboo Hall of Dongyang Guest House, 2 (Dongyang, Zhejiang)

东阳市招待所竹厅之四（浙江东阳） Bamboo Hall of Dongyang Guest House, 4 (Dongyang, Zhejiang)

东阳市招待所竹厅之一（浙江东阳） Bamboo Hall of Dongyang Guest House, 1 (Dongyang, Zhejiang)

东阳市招待所竹厅之五（浙江东阳） Bamboo Hall of Dongyang Guest House, 5 (Dongyang, Zhejiang)

东阳市招待所竹厅局部（浙江东阳） The detail of Bamboo Hall of Dongyang Guest House, 1 (Dongyang, Zhejiang)

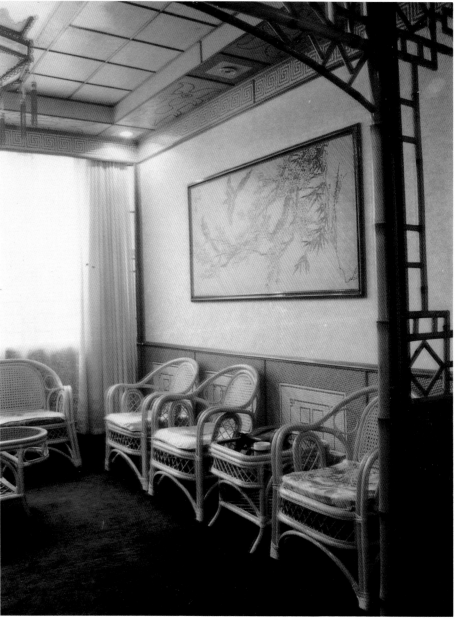

东阳市招待所竹厅之三（浙江东阳） Bamboo Hall of Dongyang Guest House, 3 (Dongyang, Zhejiang)

室内竹装潢

运用竹子天然质朴和谐的色泽,把现代的室内装潢与自然美融化为一体,用排列的圆竹作墙面,用竹丝篾片编织起来的图案作装饰,用竹枝、竹节拼镶起来的框架作门窗,再加上竹制的屏风、家具、灯具和竹雕、竹编工艺品,使室内装饰构成了竹子的艺术世界。

"Interior Bamboo Decoration"

The natural beauty of bamboo is applied ingeniously in full harmony with modern interior decoration. The walls are covered with bamboo tubes, decorated with strips woven into various patterns. The doors and windows are made of mosaicked bamboo branches and nodes. The screen, furniture, lamps and handicrafts are all made of bamboo. All these demonstrate an artistic bamboo world.

安吉宾馆天目厅(浙江安吉)
Tianmu Hall of Anji Guest House (Anji, Zhejiang)

杭州西湖竹长廊

长廊中的柱、梁、脊、瓦和栏杆,均利用竹的特殊质地经过巧拼妙镶制作,在质朴雅致的整体格调下,追求一种简约之美,充满了清新、淡逸的气韵。

"Long Bamboo Corridor on West Lake in Hangzhou"

The columns, roof beams, railings and ridge are all made of ingeniously mosaicked bamboo pieces. The simple and elegant modeling style reproduces a plain and untouched grace.

安吉宾馆五岳厅一角(浙江安吉)
Five Mountain Hall of Anji Guest House (Anji, Zhejiang)

安吉宾馆五岳厅一角(浙江安吉)
Five Mountain Hall of Anji Guest House (Anji, Zhejiang)

杭州西湖竹长廊之三(浙江杭州)
Long Bamboo Corridor on West Lake, 3 (Hangzhou

杭州西湖竹长廊之一（浙江杭州）
Long Bamboo Corridor on West Lake, 1 (Hangzhou, Zhejiang)

杭州西湖竹长廊之二（浙江杭州）
Long Bamboo Corridor on West Lake, 2 (Hangzhou, Zhejiang)

浙江绍兴柯岩风景区"蔡中郎祠"竹艺装饰正面（设计：徐华铛；制作：史舟棠、应冬春、郑兴国；策划监制：王国荣）
Front decoration of Cai Zhonglang Temple in Keyan scenic spot, Shaoxing, Zhejiang (Designer: Xu Huadang; Producers: Shi Zhoutang, Ying Dongchun, Zheng Xingguo; Managers: Lu Xinchao, Wang Guorong)

　　蔡中郎即东汉时期著名学者蔡邕，系一代才女蔡文姬之父，官至中郎将。蔡邕曾到绍兴柯岩避难，在驿馆住宿期间，看中馆中第16根椽竹，取下作笛，吹音嘹亮优雅。根据这个典故，景区用竹子来装饰"蔡中郎祠"，成为目前中国竹艺装饰中面积最大、技艺最精的场景之一，从门框到挂落，从墙裙到窗格，从画幅到对联，以至于橱柜、供桌等，均以竹子为材料，运用镶嵌、拼接、雕刻、烫烙、编织等工艺，体现出竹子的艺术，给人以清新、典雅、自然的美感，成为浙江绍兴柯岩4A级风景区的一颗装饰明珠。

Cai Zhonglang was a famous palace guard general named Cai Yi in East Han Dynastay, the father of a talented lady Cai Wenji, his official rank was "Zhonglangjiang". He came to Shaoxing for seeking asylum, staying in guesthouse, he found the sixteenth bamboo culm fit for making musical instrument, and made a sonorous flute from it. In accordance with this allusion, the Cai Zhonglang Temple is decorated with bamboo material, which is the biggest building of bamboo decoration in China. All the door frames, window frames, wall foot, table and closets are made of bamboo material with varied bamboo-processing techniques such as montage, relief, connecting and burned patterning. Consequently, aesthetic feeling of elegant, pure and fresh taste of nature is created in this building, which becomes a bright pearl in Keyan scenic spot of 4A class.

绍兴柯岩风景区"蔡中郎祠"局部
Part of Cai Zhonglang Temple, Keyan scenic spot, Shaoxing.

绍兴柯岩风景区"蔡中郎祠"局部
Part of Cai Zhonglang Temple, Keyan scenic spot, Shaoxing

绍兴柯岩"蔡中郎祠"之烙画（设计：李继渊；烙制：赵国华）

Burned picture on bamboo in Cai Zhonglang Temple (Designer: Li Jiyuan; Producer: Zhao Guohua)

绍兴柯岩风景区"蔡中郎祠"局部

Part of Cai Zhonglang Temple, Keyan scenic spot, Shaoxing.

浙江安吉叙竹庄（设计：曾伟人）
Bamboo Gathering Villa (Designer: Zeng Weiren)

浙江安吉中国竹子博物馆竹种园竹厅
Bamboo hall in bamboo species garden of Chinese Bamboo Museum, Anji, Zhejiang

上海金山宾馆内的竹门
Bamboo gate, Golden Hill Hotel, Shanghai

翻簧竹浮雕"孔雀玉兰"（浙江黄岩，罗启松作）
Engraving "Peacock and Magnolia" (Luo Qisong, Huangyan, Zhejiang)

浙江安吉中国竹子博物馆前言牌
A preface board of Chinese Bamboo Museum, Anji, Zhejiang

浙江安吉中国竹子博物馆竹装饰一角
A corner of bamboo decoration, Chinese Bamboo Museum, Anji, Zhejiang

浙江安吉中国竹子博物馆竹种园竹牌坊
Bamboo Memorial Gateway of bamboo species garden, Chinese Bamboo Museum, Anji, Zhejiang

浙江横店影视城中国竹编博物馆竹装饰一角
A corner of bamboo decoration, Chinese Bamboo Weaving Museum, Hengdian Movie Center, Zhejiang

浙江横店影视城中国竹编博物馆竹装饰一角（设计：徐华铛；制作：史济棠、应冬春等；策划监制：金顺峰、张光明）
A corner of bamboo decoration, Chinese Bamboo Weaving Museum, Hengdian Movie Center, Zhejiang (Designer: Xu Huadang, Producers: Shi Jitang, Ying Dongchun et al, Managers: Jin Shunfeng, Zhang Guangming)

浙江嵊州烟草大楼内的竹装饰
Bamboo decoration in Tobacco Building, Shengzhou, Zhejiang

竹建筑 Bamboo architecture

浙江安吉竹制小别墅（设计：曾伟人）
Villa of Bamboo, Anji, Zhejiang

傣家竹楼是傣族人民的传统民居。
This is a village of Dai people in Xishuanbana, Yunnan, China.

云南傣族、佤族和景颇族等众多少数民族都有住竹楼的传统习惯。巨竹成林环抱傣乡佤寨，翠影丛丛点缀彩云之南，一幢幢竹楼掩映在翠竹丛中，是一项珍贵的民族文化遗产，也是一项宝贵的特色旅游资源。（摄影：辉朝茂、杨宇明）
In the villages of Dai, Wa, and Jingpo pepole, beautiful bamboo houses are surrounded inside the bamboo forest. (Photos: Hui Chaomao & Yang Yuming)

佤族竹楼。
This is a village of Wa people in Ximeng county, Yunnan, China.

用竹叶盖的苦聪民居。
A house made in bamboo leaves with Kucong people in Jingping county.

勐海竹楼。
Bamboo house in Menghai county.

瑞丽的傣族竹楼。
Bamboo house in Ruili county.

由西南林学院竹类研究所和云南省竹藤产业协会负责设计和施工的中国'99昆明世界园艺博览会"竹类专题园",建成了不同风格的竹建筑,并引种竹种318种,被载入吉尼斯世界纪录。江泽慧教授视察世博竹园时欣然命笔题词,并对以著名竹类学家薛纪如教授为开拓者和奠基人的云南竹类研究丰硕成果给予高度评价。图为世博竹园的竹长廊。(摄影:辉朝茂)

This is the bamboo garden in EXPO'99. It is designed and established by the Bamboo Research Institute of Southwest Forestry College and Yunnan Bamboo and Rattan Association, has 318 species of bamboo introduced. It attracts the attention of specialists from both China and abroad and is recorded in Gynis World Record. (Photos: Hui Chaomao)

用直径达30cm的整根巨龙竹雕龙刻凤制作的牌坊,令中外来宾惊叹不已!

The gateway in the bamboo garden in EXPO'99 is carved out of four giant culms of *Dendrocalamus sinicus*.

掩映在翠竹丛中的圆形竹亭。A rotundity pavilion surrounded inside the bamboo forest.

南京玄武湖竹亭(江苏南京) Bamboo Pavilion on Xuanwu Lake in Nanjing (Nanjing, Jiangsu)

自贡彩灯博物馆竹亭(四川自贡) Bamboo Pavilion in Zigong Lantern Museum (Zigong, Sichuan)

自贡彩灯博物馆竹亭(四川自贡) Bamboo Pavilion in Zigong Lantern Museum (Zigong, Sichuan)

南京珍珠泉竹楼(江苏南京) Bamboo Pavilion on Pearl Spring in Nanjing (Nanjing, Jiangsu)

傣族竹楼 A bamboo cottage of Dai tribe

傣族竹楼
A bamboo cottage of Dai tribe

傣族竹楼
A bamboo cottage of Dai tribe

傣家竹楼（云南西双版纳）
A bamboo cottage of Dai tribe (Xishuangbanna, Yunnan)

傣家竹楼（云南西双版纳）
A bamboo cottage of Dai tribe (Xishuangbanna, Yunnan)

傣家竹建筑（云南西双版纳）
A bamboo house of Dai tribe (Xishuangbanna, Yunnan)

傣家竹亭（云南西双版纳） A bamboo pavilion of Dai tribe (Xishuangbanna, Yunnan)

瑶寨竹楼
A bamboo cottage in a Yao village

GENTLE BREEZE OVER GREEN BAMBOOS

翠竹清風

竹子挺拔、俊秀，经冬不凋，自古以来就受到人们的喜爱。中国历代文人雅士对竹子更是怀有一种特殊的感情，赞其"未出土时便有节，及凌云处尚虚心"的高尚情操。以竹为乐，爱竹成癖，以竹咏志，借竹抒怀，成为诗词、歌赋、绘画、音乐、园林等的重要题材。西晋嵇康等人有"竹林七贤"的竹林之游，王徽之有"何可一日无此君"之叹，宋代苏轼（一说文同）"宁可食无肉，不可居无竹"之赞，而清代的郑板桥把竹喻为知己，终身相伴，对竹的酷爱简直到了如痴如醉的程度。文人颂竹、吟竹者也数不胜数。如唐代李贺的"无情有恨何人见，露压烟啼千万枝"；王维的"独坐幽篁里，弹琴复长啸"；宋之问的"楚竹幽且深，半杂枫香林"；元代吴镇的"湘妃祠下竹，叶叶著秋声"；明代胡俨的"日暖风喧泪竹斑，鸣鸠拂羽树林间"；清代郑板桥的"衙斋卧听萧萧竹，疑是民间疾苦声"，一直到现代毛泽东的"斑竹一枝千滴泪，红霜万朵百重衣"等，都是脍炙人口的咏竹名句。

竹子也是中国传统绘画的重要题材，举国画者多喜绘竹，且多以水墨表现，清秀高雅。有的表现竹杆的苍劲挺拔；有的表现枝叶的孤瘦清丽；有的表现狂风下随风摇曳的"风竹"；有的则表现大雪纷飞下的"雪竹"，尽展竹之风姿，给人以不同的暇想和感受。宋代的文同，元代的李衎、赵孟頫，明代的文徵明等人都善画竹。清代画竹的名家更多，并且进入了一个新的境界，最有成就的当数"扬州八怪"之一的郑板桥。他诗、书、画皆旷世独立，人称"郑燮三绝。""四十年来画竹枝，日夜挥写夜间思，冗繁削尽留清瘦，画到生时是熟时"。郑板桥笔下的竹画，瘦劲孤高，枝枝傲霜，节节干霄，有士君子的豪气凌云。他还把竹、兰、石生动的组合在一个画面上，予以尽情的发挥，给人以笔墨之外的许多感受。他认为竹之瘦劲孤高是其神，豪迈凌云是其生，依于石而不囿于石是其节，落于色相而不滞于梗概是其品。对竹之精神认识可谓高人一等。在这本画册里，我们特选了郑板桥的《墨竹图》和《托根乱岩图》，前者用墨笔写竹四五杆，枝叶浓淡，远近显明，密而不乱，少而不疏，各有其态。后一幅在山石峭壁一侧画瘦竹几杆，高低错落有致，竹叶疏密浓淡相间，笔墨简洁明快，左上题长卷诗："咬定青山不放松，立根原在乱岩

秦汉竹简
Bamboo letters of Qin Dynasty and Han Dynasty

东汉竹简　Bamboo letter Han Dynasty

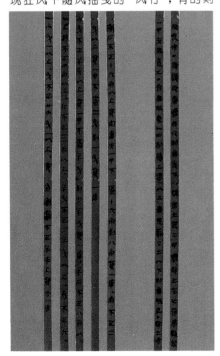

秦代竹简　Bamboo letter Qin Dynasty

窠木竹石图（元　赵孟頫）Birds' Nest, Tree, Bamboo and stone (Zhao Mengfu, Yuan Dynasty)

历代名家部分画竹精品

Masterpieces of bamboo painting by famous artists through the ages

中,千磨万击还坚劲,任尔东西南北风"。字里行间抒发出一种耐人寻味的感受,是一幅诗书画的佳作。在这本画册里,我们还选了宋代文同《墨竹图》,元代李衎《四季平安图》、赵孟頫《窠木竹石图》、柯九思《晚香高节图》,顾安《竹石图》,明代边文进《三友百禽图》、孙克弘《朱竹图》,赵备《解箨冲霄图》,清代金农《墨竹图》,元济、王原祁合作的《兰竹图》,清代石涛《灵谷探梅图》、归庄《竹石图》等。他们都是历代丹青的大家手笔,画竹都达到了精湛的程度。

中国园林中将竹列为主题进行设置的也屡见不鲜,如汉代甘泉祠宫之竹宫,宋代司马光独乐园之种竹斋。现代城市和景区内也多设有竹园,最著名的是四川成都的望江楼公园,翠竹簇簇片片,构成各个景点,是名符其实的竹林公园。其它如上海浦东的竹园新村、浙江莫干山的翠竹园、湖南洞君山的翠竹公园等,均是人们向往之地。

竹之诗,竹之画,竹之工艺,成为中国特有的文化传统,是中国独具特色的艺术瑰宝,在世界艺术之林中,有其熠熠生辉的一席之地。

四季平安图(元 李衎) Serenity of mind all the year round (Li Kan, Yuan Dynasty)

墨竹图(宋 文同)
Bamboo in black (Wen Tong, Song Dynasty)

Bamboo is tall, straight and handsome to look at. It does not wither in winter. It has been a favorite plant since ancient times. The gentlemen and men of letters in various periods of Chinese history cherished best regards to bamboos, heaping praises on its noble character, such as: "It has formed nodes before being out of soil, and has empty internodes when it is very high" (It has self-control in childhood and is humble at full manhood). To plant bamboo is a pleasure, even a hobby. Bamboo has become important subject in songs, poems, painting, music and landscape gardening, because men can express their wishes and aspirations through such a form of art.

In West Jin Dynasty Ji Kang et al were well known as "The Seven Sages in Bamboo Groves", for they often rested in bamboo groves. In EastJin Dynasty Wang Huizhi said:" How can I bahave myself without a bamboo before my presence?" In Song Dynasty Scholar Su

竹石图（元　顾安） Bamboo and stone (Gu An, Yuan Dynasty)

Shi once said:"I'd rather take a diet without meat, but there must be bamboo growing by my lodge." In Qing Dynasty scholar Zheng Banqiao deemed bamboo as an intimate friend and a life-long companion. He loved bamboo to the extreme extent. Not a few men of letters appreciated bamboo or sang a song of bamboo. For example, in Tang Dynasty poet Li He's verse was:"Who sees me having only regrets and no tender passion? But thousands of bamboo branches carry dew drops and smoke traces." Poet Wang Wei's verse was:"Sitting in deep grove I feel lonly, I play on the qin and utter a long, loud cry." Is Song Dynasty Wen Tong's verse was:"The deep grove of Chu bamboo has no fathoms. Half it is mixed with the Chinese sweet gums." In Yuan Dynasty poet Wu Zhen's verse was:"Below Queen Xiang's Temple grows a bamboo grove. Each of the rustling leaves gives an autumn note." In Ming Dynasty poet Hu Yan's verse was:"On a sunny windy day the bamboo leaves have stain of tears. Among bushes a turtledove sings and pecks its feathers." In Qing Dynasty poet Zheng Banqiao's verse was:"In my study I lie listening to the patter of rain. The sound seemed that the people in distress do complain." In modern times the late Chairman Mao Zedong's verse was: One branch of mottled bamboo has thousand drops of tears. Myriad red leaves tinged by frost are like hundred colored dresses." All these are oft-quoted and widely loved verses on bamboo.

Bamboo is an important subject of traditional Chinese painting. Most of the Chinese painting pursuers like to paint bamboo. The bamboos mostly drawn in black ink are graceful and elegant. Some culms are robust, tall and straight. Some branches with leaves are lean but handsome. Some bamboos swing in the sweeping storm. Some weigh down under the incessant heavy snow. The bamboo posture under various circumstances catches the fancy of people and makes different impression on them. Scholar Wen Tong and scholar Li Yan and Zhao Mengfu in Yuan Dynasty, and scholar Wen Zhengming in Ming Dynasty were all good at painting bamboo. In Qing Dynasty there scholars good at painting bamboo, and their technique had raised to a new level. Most achievements were scored by one of the Eight Eccentrics, Zheng Banqiao. His poetry, calligraphy and painting were independent and unique, and were known as "Three Consummate Skills of Zheng Qian." One of his poems on bamboo was: "For forty years bamboo branches I drew. In daytime I wielded brush, in night I did conceive. With redundance cut, the leanness was conducive. The Skill was ripe when I created something new." The bamboo he drew were lean, robust, lonely and elegant. Each branch was tolerant to frost, and each internode shot up towards sky like a scholar or gentleman cherishing a noble aim. In his drawing he combined bamboo with orchid and stone, bringing them all into prominence, and conveying more impressions than brush and ink alone could do, in his conception, the leanness, robustness, loneliness and elegance were bamboo's spirit; boldness, generosity and high aspiration were its life. Standing by a stone but being not limited by it, it maintained its independent existence. It had a true charm, not a superficial one, which was the symbol of its high moral character. His knowledge of bamboo spirit was above that of his contemporaries. This album contains two pictures from Zheng. One is "bamboo draw in black ink" and the other is "Striking roots among cracked rocks." In the former there were 4,5 culms drawn with Chinese brush, with branches and leaves in various shades and in a good perspective. The foliage is dense, but not irregular. The culm were few, but not sparse. Each has its own particular features. In the latter, by the side of a cliff there were a few lean bamboos. They varied in plant height, but as a whole they looked quite well. Some branches were covered with dense foliage and some had only sparse leaves. The color varied in shades. The picture afforded an impression of simplicity and brevity characteristic of a lucid and lively style. On the left upper corner is a stanza of long poem: "It makes a tight bite of the green hill. Its roots are established among cracked rocks. After repeated frictions it's firm still, braving any wing signaled by the weather cocks.

It is interesting and thought-provoking. This album contains other famous paintings such as:" Bamboo in black" by Wen Tong in Song Dynasty. "Serenity of mind all the year round"by Li Kan. "Birds' nest, tree, bamboo and stone" by Zhao Mengfu," Autumn fragrance and erect culms" by Ke Jiusi, "Bamboo and stone"by Gu An in Yuan Dynasty," Three friends and numerus birds" by Bian WenJin and "Bamboo in red" by Sun Kehong in Ming Dynasty, "bamboo in black" by Jin Nong. "Orchid and bamboo" by Yuan Ji and Wang Yuanqi. "Visit plums in Linggu dale" by Shi Tao, and "Baboo and stone" by Gui Zhuang in Qing Dynasty. All of the authors were masters of painting of their times, they were highly skilled in painting bamboo.

In the landscape gardening in China there are many exaqmples of setting with bamboo as the main theme, such as the Bamboo Palace in Ganquan Temple in Han Dynasty, the Bamboo Planting Study in the picture of Individual Pleasure by Si Maguang in Song Dynastey. In

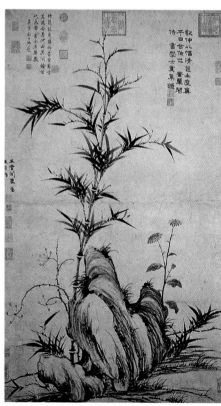

晚香高节图（元　柯九思） Autumn fragrance and erect culms (Ke Jiusi, Yuan Dynasty)

解箨冲霄图（明 赵备）Breaking shells and pointing to clouds (Zhao Bei, Ming Dynasty)

朱竹图（明 孙克弘） Bamboo in red (Sun Kehong, Ming Dynasty)

三友百禽图（明 边文进） Three friends and numerus birds (Bian Wenjin, Ming Dynasty)

Bamboo Park on Dongjun Hill in Hunan province, all drawing much attention of people.

The poems on bamboo, pictures of bamboo and bamboo craftsman ship are particular objects in the traditional Chinese culture and are valuables works of art characteristic of China, occupying a magnificent position in the treasury of world art.

modern cities and scenic spots there are often bamboo gardens. The most famous one is Wangjianglou Park in Chengdu, Sichuan province, in which there are many groves, forming separate scenic spots. It is a really bamboo park true to its name. Other bamboo parks such as New Bamboo Park Village at Pudong Area in Shanghai, Green Bamboo Park in Mogan Hill in Zhejiang province, and Green

竹石图（清 归庄） Bamboo and stone (Gui Zhuang, Qing Dynasty)

173

兰竹图（清 元济，王原祁合作） Orchid and bamboo (Yuan Ji and Wang Yuanqi, Qing Dynasty)

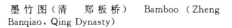

墨竹图（清 郑板桥） Bamboo (Zheng Banqiao, Qing Dynasty)

托根乱岩图（清 郑板桥） Growing on wild rocks (Zheng Banqiao, Qing Dynasty)

灵谷探梅图（清 石涛） Visiting the plum in a magic valley (Shi Tao, Qing Dynasty)

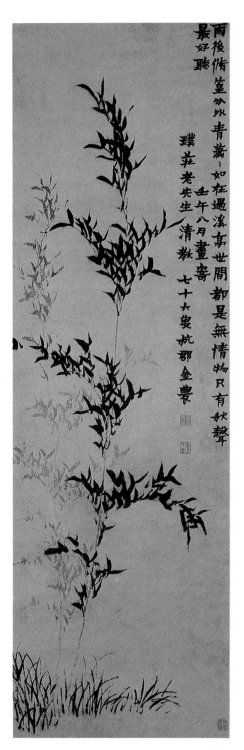

墨竹图（清　金农）Bamboo in black (Jin Nong, Qing Dynasty)

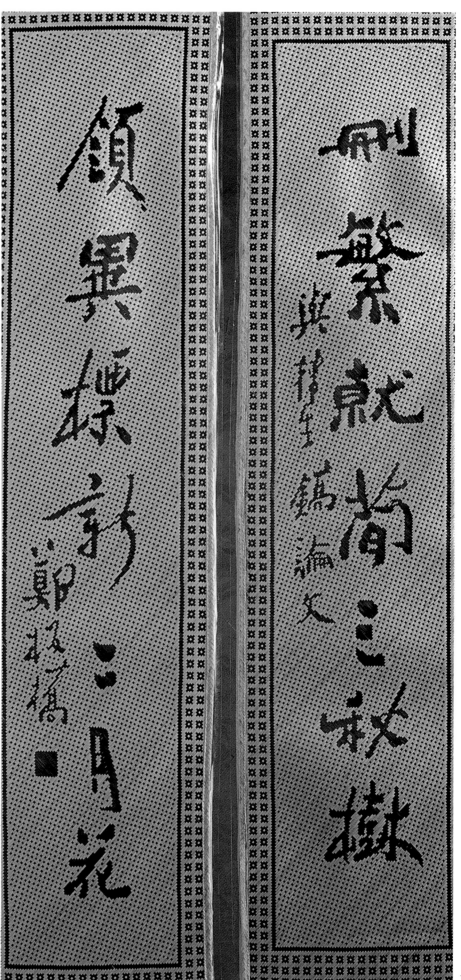

郑板桥书法（四川渠县）
Calligraphy by Zheng Banqiao (Quxian, Sichuan)

竹编郑板桥竹画（四川渠县）
Painting by Zheng Banqiao "Bamboo" (Quxian, Sichuan)

郑板桥书法（四川渠县）　Calligraphy by Zheng Banqiao (Quxian, Sichuan)

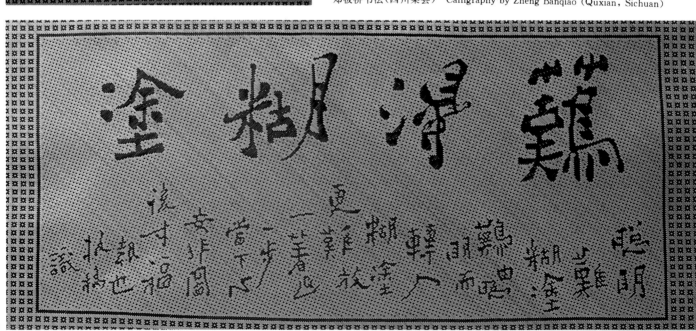